Ben Stacy Jerrik (Ed.)

Dual Polarization Interferometry

Ben Stacy Jerrik (Ed.)

Dual Polarization Interferometry

Waveguide (optics), Evanescent wave, Laser

Part Press

Publisher:
Part Press is a trademark of
International Book Market Service Ltd., 17 Rue Meldrum, Beau Bassin, 1713-01 Mauritius
Email: info@bookmarketservice.com
Website: www.bookmarketservice.com

Published in 2012

Printed in: U.S.A., U.K., Germany. This book was not produced in Mauritius.

ISBN: 978-620-1-64789-3

Contents

Articles

References

Dual_polarization_interferometry

Dual polarization interferometry (**DPI**) is an analytical technique that can probe molecular scale layers adsorbed to the surface of a waveguide by using the evanescent wave of a laser beam confined to the waveguide. Typically used to measure the conformational change in proteins or other biomolecules as they function (referred to as the conformation activity relationship).

DPI[1] focuses laser light into two waveguides. One of these functions as the "sensing" waveguide having an exposed surface while the second one functions to maintain a reference beam. A two-dimensional interference pattern is formed in the far field by combining the light passing through the two waveguides. The DPI technique rotates the polarization of the laser to alternately excite two polarization modes of the waveguides. Measurement of the interferogram for both polarizations allows both the refractive index and the thickness of the adsorbed layer to be calculated. The polarization can be switched rapidly, allowing real-time measurements of chemical reactions taking place on a chip surface in a flow-through system. These measurements can be used to infer conformational information about the molecular interactions taking place as the molecule size (from the layer thickness) and the fold density (from the RI) change. DPI is typically used to characterise biochemical interactions by quantifying any conformational change at the same time as measuring reaction rates, affinities and thermodynamics.

The technique is quantitative and real-time (10 Hz) with a dimensional resolution of 0.01 nm.[2]

A novel application for Dual Polarization Interferometry emerged in 2008 where the intensity of light passing through the waveguide is extinguished in the presence of crystal growth. This has allowed the very earliest stages in protein crystal nucleation to be monitored.[3] Later versions of Dual Polarization Interferometers also have the capability to quantify the order and disruption in birefringent thin films.[4] This has been used, for example, to study the formation of lipid bilayers and their interaction with membrane proteins.[5] [6] In December 2011, the interferometer system was announced as a discontinued product line although existing users were to be supported until December 2016. A web based discussion page [7] was opened in April 2012 for existing and potential users to network and provide information.

References

[1] Cross, G; Reeves, AA; Brand, S; Popplewell, JF; Peel, LL; Swann, MJ; Freeman, NJ (2003). "A new quantitative optical biosensor for protein characterisation". *Biosensors and Bioelectronics* **19** (4): 383. doi:10.1016/S0956-5663(03)00203-3. PMID 14615097.

[2] Swann, MJ; Freeman, NJ; Cross, GH (2007). "Dual Polarization Interferometry: A Real-Time Optical Technique for Measuring (Bio)Molecular Orientation, Structure and Function at the Solid/Liquid Interface". In Marks, R.S; Lowe, C.R.; Cullen, D.C.; Weetall, H.H.; Karube, I.. *Handbook of Biosensors and Biochips*. **Vol. 1**. Wiley. pt. 4, ch. 33, pp. 549–568. ISBN 978-0-470-01905-4.

[3] Boudjemline, A; Clarke, DT; Freeman, NJ; Nicholson, JM; Jones, GR (2008). "Early stages of protein crystallization as revealed by emerging optical waveguide technology". *Journal of Applied Crystallography* **41**: 523. doi:10.1107/S0021889808005098.

[4] Mashaghi, A; Swann, M; Popplewell, J; Textor, M; Reimhult, E (2008). "Optical Anisotropy of Supported Lipid Structures Probed by Waveguide Spectroscopy and Its Application to Study of Supported Lipid Bilayer Formation Kinetics". *Analytical Chemistry* **80** (10): 3666. doi:10.1021/ac800027s. PMID 18422336.

[5] Sanghera, N; Swann, MJ; Ronan, G; Pinheiro, TJ (2009). "Insight into early events in the aggregation of the prion protein on lipid membranes". *Biochimica et Biophysica Acta* **1788** (10): 2245–51. doi:10.1016/j.bbamem.2009.08.005. PMID 19703409.

[6] Lee, TH; Heng, C; Swann, MJ; Gehman, JD; Separovic, F; Aguilar, MI (2010). "Real-time quantitative analysis of lipid disordering by aurein 1.2 during membrane adsorption, destabilisation and lysis". *Biochimica et Biophysica Acta* **1798** (10): 1977–86. doi:10.1016/j.bbamem.2010.06.023. PMID 20599687.

[7] http://dpi-insider.proboards.com/

Further reading

- Cross, GH; Ren, Y; Freeman, NJ (1999). "Young's fringes from vertically integrated slab waveguides: Applications to humidity sensing". *Journal of Applied Physics* **86**: 6483. Bibcode 1999JAP....86.6483C. doi:10.1063/1.371712.
- Cross, G (2003). "A new quantitative optical biosensor for protein characterisation". *Biosensors and Bioelectronics* **19**: 383. doi:10.1016/S0956-5663(03)00203-3. PMID 14615097.
- Freeman, NJ; Peel, LL; Swann, MJ; Cross, GH; Reeves, A; Brand, S; Lu, JR (2004). "Real time, high resolution studies of protein adsorption and structure at the solid–liquid interface using dual polarization interferometry". *Journal of Physics: Condensed Matter* **16**: S2493. Bibcode 2004JPCM...16S2493F. doi:10.1088/0953-8984/16/26/023.
- Khan, TR; Grandin, HM; Mashaghi, A; Textor, M; Reimhult, E; Reviakine, I (2008). "Lipid redistribution in phosphatidylserine-containing vesicles adsorbing on titania.". *Biointerphases* **3** (2): FA90. doi:10.1116/1.2912098. PMID 20408675.

Waveguide_(optics)

An **optical waveguide** is a physical structure that guides electromagnetic waves in the optical spectrum. Common types of optical waveguides include optical fiber and rectangular waveguides.

Optical waveguides are used as components in integrated optical circuits or as the transmission medium in local and long haul optical communication systems.

Optical waveguides can be classified according to their geometry (planar, strip, or fiber waveguides), mode structure (single-mode, multi-mode), refractive index distribution (step or gradient index) and material (glass, polymer, semiconductor).

Dielectric slab waveguide

Practical rectangular-geometry optical waveguides are most easily understood as variants of the simple dielectric **slab waveguide** RWvD1984, also called **planar waveguide**[1] . The slab waveguide consists of three layers of materials with different dielectric constants, extending infinitely in the directions parallel to their interfaces.

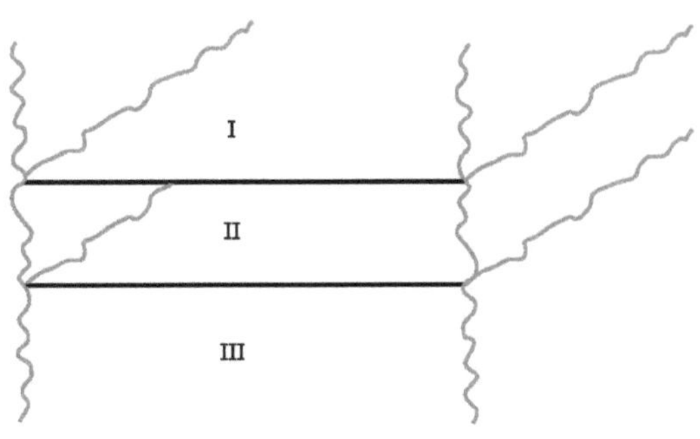

A dielectric slab waveguide consists of three dielectric layers with different refractive indices.

Light may be confined in the middle layer by total internal reflection. This occurs only if the dielectric index of the middle layer is larger than that of the surrounding layers. In practice slab waveguides are not infinite in the direction parallel to the interface, but if the typical size of the interfaces is much much larger than the depth of the layer, the slab waveguide model will be an excellent approximation. Guided modes of a slab waveguide can not be excited by light incident from the top or bottom interfaces. Light must be injected with a lens from the side into the middle layer. Alternatively a coupling element may be used to couple light into the waveguide, such as a grating coupler or prism coupler.

One model of guided modes is that of a planewave reflected back and forth between the two interfaces of the middle layer, at an angle of incidence between the propagation direction of the light and the *normal*, or perpendicular direction, to the material interface is greater than the *critical angle*. The critical angle depends on the index of refraction of the materials, which may vary depending on the wavelength of the light. Such propagation will result in a guided mode only at a discrete set of angles where the reflected planewave does not destructively interfere with itself.

This structure confines electromagnetic waves only in one direction, and therefore it has little practical application. Structures that may be approximated as slab waveguides do, however, sometimes occur as incidental structures in other devices.

Two-dimensional waveguides

Strip waveguides

A **strip waveguide** is basically a strip of the guiding layer confined between cladding layers. The simplest case is a **rectangular waveguide**, which is formed when the guiding layer of the slab waveguide is restricted in both transverse directions rather than just one. Rectangular waveguides are used in integrated optical circuits, and in laser diodes. They are commonly used as the basis of such optical components as Mach-Zehnder interferometers and wavelength division multiplexers. The cavities of laser diodes are frequently constructed as rectangular optical waveguides. Optical waveguides with rectangular geometry are produced by a variety of means, usually by a planar process.

The field distribution in rectangular waveguide cannot be solved analytically, however approximate solution methods, such as Marcatili's method, are known.

Rib waveguides

A **rib waveguide** is a waveguide in which the guiding layer basically consists of the slab with a strip (or several strips) superimposed onto it. Rib waveguides also provide confinement of the wave in two dimensions.

Optical fiber

Optical fiber is typically a circular cross-section *dielectric waveguide* consisting of a dielectric material surrounded by another dielectric material with a lower refractive index. Optical fibers are most commonly made from silica glass, however other glass materials are used for certain applications and plastic optical fiber can be used for short-distance applications.

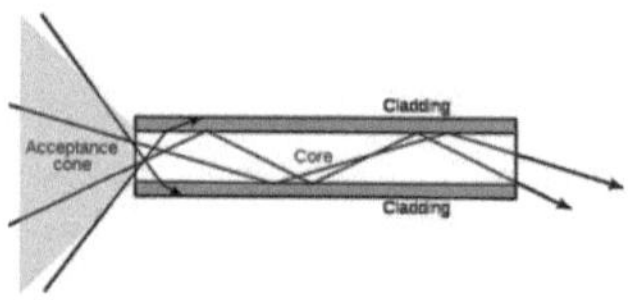

The propagation of light through a multi-mode optical fiber.

See also

- ARROW waveguide
- Cutoff wavelength
- Dielectric constant
- Digital planar holography
- Electromagnetic radiation
- Erbium-doped waveguide amplifier
- Equilibrium mode distribution
- Leaky mode
- Lightguide display
- Transmission medium
- Waveguide
- Waveguide (electromagnetism)
- Photonic crystal fiber
- Photonic crystal
- Prism coupler
- Zero-mode waveguide

Notes

1. Ramo, Simon, John R. Whinnery, and Theodore van Duzer, *Fields and Waves in Communications Electronics*, 2 ed., John Wiley and Sons, New York, 1984.

[1] "Silicon Photonics", by Graham T. Reed, Andrew P. Knights

External links

- AdvR - Rubidium doped waveguides in potassium titanyl phosphate (KTP) (http://www.advr-inc.com/waveguide.html)

Evanescent_wave

An **evanescent wave** is a near-field standing wave with an intensity that exhibits exponential decay with distance from the boundary at which the wave was formed. Evanescent waves are a general property of wave-equations, and can in principle occur in any context to which a wave-equation applies. They are formed at the boundary between two media with different wave motion properties, and are most intense within one third of a wavelength from the surface of formation. In particular, evanescent waves can occur in the contexts of optics and other forms of electromagnetic radiation, acoustics, quantum mechanics, and "waves on strings".[1] [2]

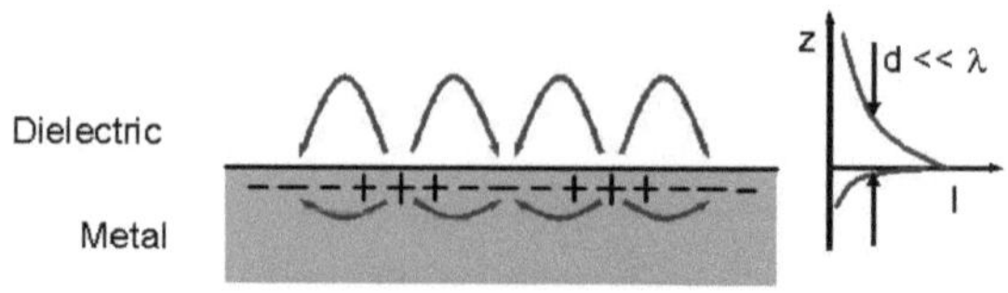

Schematic representation of evanescent waves propagating along a metal-dielectric interface. The charge density oscillations, when associated with electromagnetic fields, are called surface plasmon-polariton waves. The exponential dependence of the electromagnetic field intensity on the distance away from the interface is shown on the right. These waves can be excited very efficiently with light in the visible range of the electromagnetic spectrum.

Evanescent wave applications

In optics and acoustics, evanescent waves are formed when waves traveling in a medium undergo total internal reflection at its boundary because they strike it at an angle greater than the so-called *critical angle*.[1] [2] The physical explanation for the existence of the evanescent wave is that the electric and magnetic fields (or pressure gradients, in the case of acoustical waves) cannot be discontinuous at a boundary, as would be the case if there was no evanescent wave field. In quantum mechanics, the physical explanation is exactly analogous—the Schrödinger wave-function representing particle motion normal to the boundary cannot be discontinuous at the boundary.

Electromagnetic evanescent waves have been used to exert optical radiation pressure on small particles in order to trap them for experimentation, or to cool them to very low temperatures, and to illuminate very small objects such as biological cells for microscopy (as in the total internal reflection fluorescence microscope). The evanescent wave from an optical fiber can be used in a gas sensor, and evanescent waves figure in the infrared spectroscopy technique known as attenuated total reflectance.

In electrical engineering, evanescent waves are found in the near-field region within one third of a wavelength of any radio antenna. During normal operation, an antenna emits electromagnetic fields into the surrounding nearfield region, and a portion of the field energy is reabsorbed, while the remainder is radiated as EM waves.

In quantum mechanics, the evanescent-wave solutions of the Schrödinger equation give rise to the phenomenon of wave-mechanical tunneling.

In microscopy, systems that capture the information contained in evanescent waves can be used to create super-resolution images. Matter radiates both propagating and evanescent electromagnetic waves. Conventional optical systems capture only the information in the propagating waves and hence are subject to the diffraction limit. Systems that capture the information contained in evanescent waves, such as the superlens and near field scanning optical microscopy, can overcome the diffraction limit; however these systems are then limited by the system's ability to accurately capture the evanescent waves.[3] The limitation on their resolution is given by

,

where is the maximum wave vector that can be resolved, is the distance between the object and the sensor, and is a measure of the quality of the sensor.

More generally, practical applications of evanescent waves can be classified in the following way:

1. Those in which the energy associated with the wave is used to excite some other phenomenon within the region of space where the original traveling wave becomes evanescent (for example, as in the total internal reflection fluorescence microscope)
2. Those in which the evanescent wave couples two media in which traveling waves are allowed, and hence permits the transfer of energy or a particle between the media (depending on the wave equation in use), even though no traveling-wave solutions are allowed in the region of space between the two media. An example of this is so-called *wave-mechanical tunnelling*, and is known generally as *evanescent wave coupling*.

Total internal reflection of light

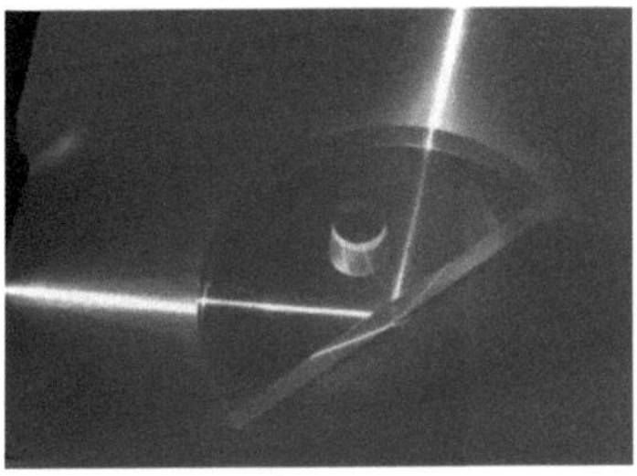

Total internal reflection

For example, consider total internal reflection in two dimensions, with the interface between the media lying on the x axis, the normal along y, and the polarization along z. One might naively expect that for angles leading to total internal reflection, the solution would consist of an incident wave and a reflected wave, with no transmitted wave at all, but there is no such solution that obeys Maxwell's equations. Maxwell's equations in a dielectric medium impose a boundary condition of continuity for the components of the fields $E_{||}$, $H_{||}$, D_y, and B_y. For the polarization considered in this example, the conditions on $E_{||}$ and B_y are satisfied if the reflected wave has the same amplitude as the incident one, because these components of the incident and reflected waves superimpose destructively. Their H_x components, however, superimpose constructively, so there can be no solution without a non-vanishing transmitted wave. The transmitted wave cannot, however, be a sinusoidal wave, since it would then transport energy away from the boundary, but since the incident and reflected waves have equal energy, this would violate conservation of energy. We therefore conclude that the transmitted wave must be a non-vanishing solution to Maxwell's equations that is not a traveling wave, and the only such solutions in a dielectric are those that decay exponentially: evanescent waves.

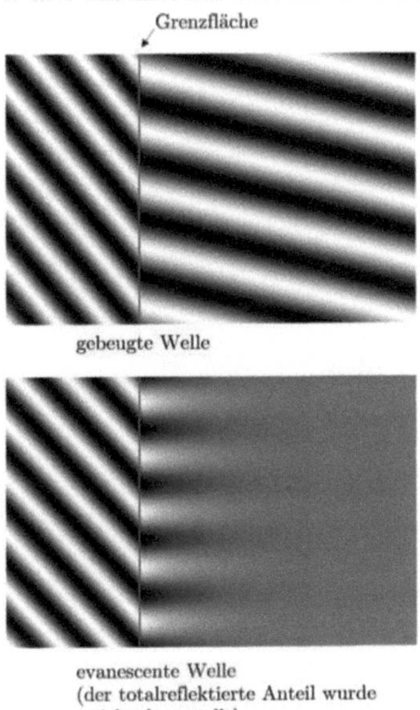

Top to bottom: representation of a refracted incident wave and an evanescent wave at an interface.

Mathematically, evanescent waves can be characterized by a wave vector where one or more of the vector's components has an imaginary value. Because the vector has imaginary components, it may have a magnitude that is less than its real components. If the angle of incidence exceeds the critical angle, then the wave vector of the transmitted wave has the form

which represents an evanescent wave because the y component is imaginary. (Here α and β are real and i represents the imaginary unit.)

For example, if the polarization is perpendicular to the plane of incidence, then the electric field of any of the waves (incident, reflected, or transmitted) can be expressed as where is the unit vector in the z direction.

Substituting the evanescent form of the wave vector **k** (as given above), we find for the transmitted wave:

where α is the *attenuation constant* and β is the *propagation constant.*

Evanescent-wave coupling

In optics, *evanescent-wave coupling* is a process by which electromagnetic waves are transmitted from one medium to another by means of the evanescent, exponentially decaying electromagnetic field.

Coupling is usually accomplished by placing two or more electromagnetic elements such as optical waveguides close together so that the evanescent field generated by one element does not decay much before it reaches the other element. With waveguides, if the receiving waveguide can support modes of the appropriate frequency, the evanescent field gives rise to propagating-wave modes, thereby connecting (or coupling) the wave from one waveguide to the next.

Evanescent-wave coupling is fundamentally identical to near field interaction in electromagnetic field theory. Depending on the impedance of the radiating source element, the evanescent wave is either predominantly electric (capacitive) or magnetic (inductive), unlike in the far field where these components of the wave eventually reach the ratio of the impedance of free space and the wave propagates radiatively. The evanescent wave coupling takes place in the non-radiative field near each medium and as such is always associated with matter; i.e., with the induced currents and charges within a partially reflecting surface. This coupling is directly analogous to the coupling between the primary and secondary coils of a transformer, or between the two plates of a capacitor. Mathematically, the process is the same as that of quantum tunneling, except with electromagnetic waves instead of quantum-mechanical wavefunctions.

Applications

- Evanescent wave coupling is commonly used in photonic and nanophotonic devices as waveguide sensors.
- Evanescent wave coupling is used to excite, for example, dielectric microsphere resonators.
- A typical application is resonant energy transfer, useful, for instance, for charging electronic gadgets without wires. A particular implementation of this is WiTricity; the same idea is also used in some Tesla coils.
- Evanescent coupling, as near field interaction, is one of the concerns in electromagnetic compatibility.
- Evanescent wave coupling plays a major role in the theoretical explanation of extraordinary optical transmission.[4]

See also

- Coupling (electronics)
- Electromagnetic wave
- Quantum tunneling
- Resonant energy transfer
- Snell's law
- Total internal reflection
- Total internal reflection fluorescence microscope
- Waveguide

References

- Karalis, Aristeidis; J.D. Joannopoulos, Marin Soljačić (February 2007). "Efficient wireless non-radiative mid-range energy transfer". *Annals of Physics* **323**: 34. arXiv:physics/0611063v2. Bibcode 2008AnPhy.323...34K. doi:10.1016/j.aop.2007.04.017.
- "'Evanescent coupling' could power gadgets wirelessly", Celeste Biever, *NewScientist.com*, 15 November 2006 [5]
- Wireless energy could power consumer, industrial electronics [6] - MIT press release

[1] Tineke Thio (2006). "A Bright Future for Subwavelength Light Sources". *American Scientist* (American Scientist) **94** (1): 40–47. doi:10.1511/2006.1.40.

[2] Marston, Philip L.; Matula, T.J. (May 2002). "Scattering of acoustic evanescent waves...". *Journal of the Acoustical Society of America* **111** (5): 2378. Bibcode 2002ASAJ..111.2378M.

[3] Neice, A., "Methods and Limitations of Subwavelength Imaging", Advances in Imaging and Electron Physics, Vol. 163, July 2010

[4] Z. Y. Fan, L. Zhan, X. Hu, and Y. X. Xia, "Critical process of extraordinary optical transmission through periodic subwavelength hole array: Hole-assisted evanescent-field coupling," Optics Communications 281, 5467-5471 (2008).

[5] http://www.newscientist.com/article/dn10575-evanescent-coupling-could-power-gadgets-wirelessly.html

[6] http://web.mit.edu/newsoffice/2006/wireless.html

External links

- Evanescent waves (http://www.andrew.cmu.edu/user/dcprieve/Evanescent waves.htm)
- An animation from a product manufacturer that makes use of evanescent fields (http://www.farfieldgroup.shared.hosting.zen.co.uk/technology_dpi.asp)
- Evanescent and propagating waves animation on Youtube.com (http://www.youtube.com/watch?v=UhtczKSO-Us)

Laser

A **laser** is a device that emits light (electromagnetic radiation) through a process of optical amplification based on the stimulated emission of photons. The term "laser" originated as an acronym for *Light Amplification by Stimulated Emission of Radiation*.[1] [2] The emitted laser light is notable for its high degree of spatial and temporal coherence, unattainable using other technologies.

United States Air Force laser experiment

Spatial coherence typically is expressed through the output being a narrow beam which is diffraction-limited, often a so-called "pencil beam." Laser beams can be focused to very tiny spots, achieving a very high irradiance. Or they can be launched into a beam of very low divergence in order to concentrate their power at a large distance.

Temporal (or longitudinal) coherence implies a polarized wave at a single frequency whose phase is correlated over a relatively large distance (the coherence length) along the beam.[3] A beam produced by a thermal or other incoherent light source has an instantaneous amplitude and phase which vary randomly with respect to time and position, and thus a very short coherence length.

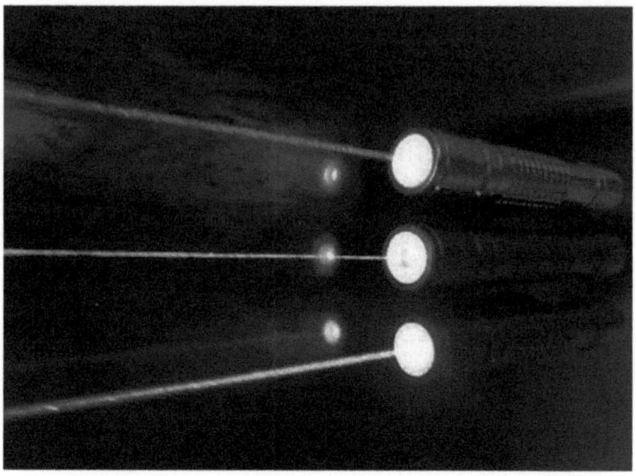

Red (635 nm), green (532 nm), and blue-violet (445 nm) lasers

Most so-called "single wavelength" lasers actually produce radiation in several *modes* having slightly different frequencies (wavelengths), often not in a single polarization. And although temporal coherence implies monochromaticity, there are even lasers that emit a broad spectrum of light, or emit different wavelengths of light simultaneously. There are some lasers which are not single spatial mode and consequently their light beams diverge more than required by the diffraction limit. However all such devices are classified as "lasers" based on their method of producing that light: stimulated emission. Lasers are employed in applications where light of the required spatial or temporal coherence could not be produced using simpler technologies.

Terminology

The word *laser* started as an acronym for "light amplification by stimulated emission of radiation"; in modern usage "light" broadly denotes electromagnetic radiation of any frequency, not only visible light, hence *infrared laser*, *ultraviolet laser*, *X-ray laser*, and so on. Because the microwave predecessor of the laser, the maser, was developed first, devices of this sort operating at microwave and radio frequencies are referred to as "masers" rather than "microwave lasers" or "radio lasers". In the early technical literature, especially at Bell Telephone Laboratories, the laser was called an **optical maser**; this term is now obsolete.[4]

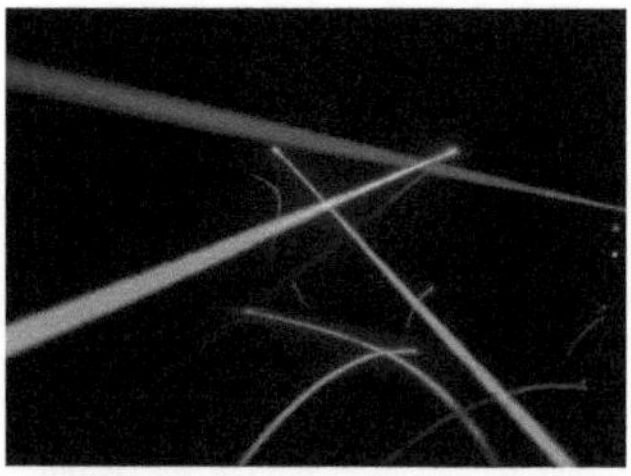

Laser beams in fog, reflected on a car windshield

A laser which produces light by itself is technically an optical oscillator rather than an optical amplifier as suggested by the acronym. It has been humorously noted that the acronym LOSER, for "light oscillation by stimulated emission of radiation," would have been more correct.[5] With the widespread use of the original acronym as a common noun, actual optical amplifiers have come to be referred to as "laser amplifiers", notwithstanding the apparent redundancy in that designation.

The back-formed verb *to lase* is frequently used in the field, meaning "to produce laser light,"[6] especially in reference to the gain medium of a laser; when a laser is operating it is said to be "lasing." Further use of the words *laser* and *maser* in an extended sense, not referring to laser technology or devices, can be seen in usages such as *astrophysical maser* and *atom laser*.

Design

A laser consists of a gain medium, a mechanism to supply energy to it, and something to provide optical feedback.[7] The gain medium is a material with properties that allow it to amplify light by stimulated emission. Light of a specific wavelength that passes through the gain medium is amplified (increases in power).

For the gain medium to amplify light, it needs to be supplied with energy. This process is called pumping. The energy is typically supplied as an electrical current, or as light at a different wavelength. Pump light may be provided by a flash lamp or by another laser.

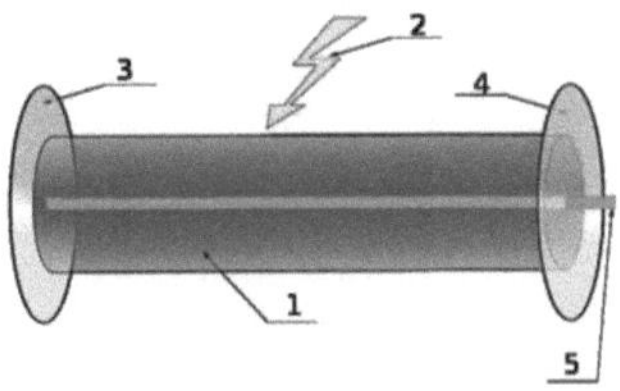

Components of a typical laser:
1. Gain medium
2. Laser pumping energy
3. High reflector
4. Output coupler
5. Laser beam

The most common type of laser uses feedback from an optical cavity—a pair of mirrors on either end of the gain medium. Light bounces back and forth between the mirrors, passing through the gain medium and being amplified each time. Typically one of the two mirrors, the output coupler, is partially transparent. Some of the light escapes through this mirror. Depending on the design of the cavity (whether the mirrors are flat or curved), the light coming out of the laser may spread out or form a narrow beam. This type of device is sometimes called a *laser oscillator* in analogy to electronic oscillators, in which an electronic amplifier receives electrical feedback that causes it to produce a signal.

Most practical lasers contain additional elements that affect properties of the emitted light such as the polarization, the wavelength, and the shape of the beam.

Laser physics

Electrons and how they interact with electromagnetic fields are important in our understanding of chemistry and physics.

Stimulated emission

In the classical view, the energy of an electron orbiting an atomic nucleus is larger for orbits further from the nucleus of an atom. However, quantum mechanical effects force electrons to take on discrete positions in orbitals. Thus, electrons are found in specific energy levels of an atom, two of which are shown below:

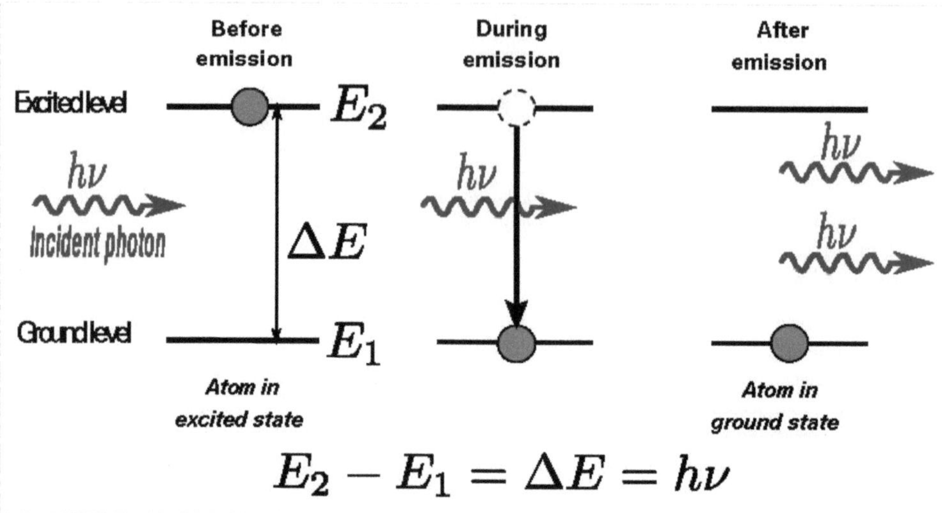

When an electron absorbs energy either from light (photons) or heat (phonons), it receives that incident quanta of energy. But transitions are only allowed in between discrete energy levels such as the two shown above. This leads to emission lines and absorption lines.

When an electron is excited from a lower to a higher energy level, it will not stay that way forever. An electron in an excited state may decay to a lower energy state which is not occupied, according to a particular time constant characterizing that transition. When such an electron decays without external influence, emitting a photon, that is called "spontaneous emission". The phase associated with the photon that is emitted is random. A material with many atoms in such an excited state may thus result in radiation which is very spectrally limited (centered around one wavelength of light), but the individual photons would have no common phase relationship and would emanate in random directions. This is the mechanism of fluorescence and thermal emission.

An external electromagnetic field at a frequency associated with a transition can affect the quantum mechanical state of the atom. As the electron in the atom makes a transition between two stationary states (neither of which shows a dipole field), it enters a transition state which does have a dipole field, and which acts like a small electric dipole, and this dipole oscillates at a characteristic frequency. In response to the external electric field at this frequency, the probability of the atom entering this transition state is greatly increased. Thus, the rate of transitions between two stationary states is enhanced beyond that due to spontaneous emission. Such a transition to the higher state is called absorption, and it destroys an incident photon (the photon's energy goes into powering the increased energy of the higher state). A transition from the higher to a lower energy state, however, produces an additional photon; this is the process of **stimulated emission**.

Gain medium and cavity

The gain medium is excited by an external source of energy into an excited state. In most lasers this medium consists of population of atoms which have been excited into such a state by means of an outside light source, or a electrical field which supplies energy for atoms to absorb and be transformed into their excited states.

The gain medium of a laser is normally a material of controlled purity, size, concentration, and shape, which amplifies the beam by the process of stimulated emission described above. This material can be of any state: gas, liquid, solid, or plasma. The gain medium absorbs pump energy, which raises some electrons into higher-energy ("excited") quantum states. Particles can interact with light by either absorbing or emitting photons. Emission can be spontaneous or stimulated. In the latter case, the photon is emitted in the same direction as the light that is passing by. When the number of particles in one excited state exceeds the number of particles in some lower-energy state, population inversion is achieved and the amount of stimulated emission due to light that passes through is larger than the amount of absorption. Hence, the light is amplified. By itself, this makes an optical amplifier. When an optical amplifier is placed inside a resonant optical cavity, one obtains a laser oscillator.[8]

A helium-neon laser demonstration at the Kastler-Brossel Laboratory at Univ. Paris 6. The pink-orange glow running through the center of the tube is from the electric discharge which produces incoherent light, just as in a neon tube. This glowing plasma is excited and then acts as the gain medium through which the internal beam passes, as it is reflected between the two mirrors. Laser radiation output through the front mirror can be seen to produce a tiny (about 1mm in diameter) intense spot on the screen, to the right. Although it is a deep and pure red color, spots of laser light are so intense that cameras are typically overexposed and distort their color.

In a few situations it is possible to obtain lasing with only a single pass of EM radiation through the gain medium, and this produces a laser beam without any need for a resonant or reflective cavity (see for example nitrogen laser).[9] Thus, reflection in a resonant cavity is usually required for a laser, but is not absolutely necessary.

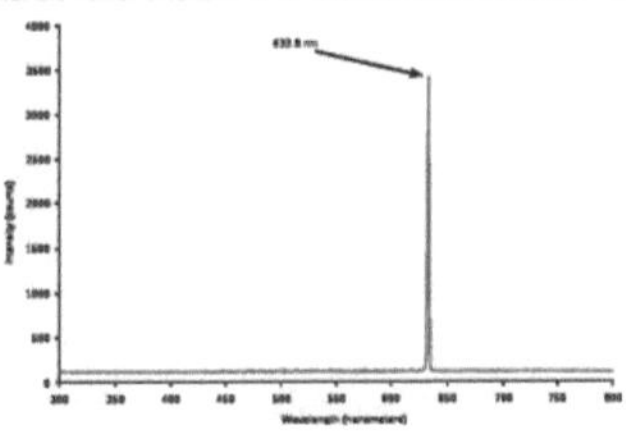

Spectrum of a helium neon laser illustrating its very high spectral purity (limited by the measuring apparatus). The .002 nm bandwidth of the lasing medium is well over 10,000 times narrower than the spectral width of a light-emitting diode (whose spectrum is shown **here** for comparison), with the bandwidth of a single longitudinal mode being much narrower still.

The optical resonator is sometimes referred to as an "optical cavity", but this is a misnomer: lasers use open resonators as opposed to the literal cavity that would be employed at microwave frequencies in a maser. The resonator typically consists of two mirrors between which a coherent beam of light travels in both directions, reflecting back on itself so that an average photon will pass through the gain medium repeatedly before it is emitted from the output aperture or lost to diffraction or absorption. If the gain (amplification) in the medium is larger than the resonator losses, then the power of the recirculating light can rise exponentially. But each stimulated emission event returns an atom from its excited state to the ground state, reducing the gain of the medium. With increasing beam power the net gain (gain times loss) reduces to unity and the gain medium is said to be saturated. In a continuous wave (CW) laser, the balance of pump power against gain saturation and cavity losses produces an equilibrium value of the laser power inside the cavity; this equilibrium determines the operating point of the laser. If the applied pump power is too small, the gain will never be sufficient to overcome the resonator losses, and laser light will not be produced. The minimum pump power needed to begin laser action is called the *lasing threshold*. The gain medium will amplify any photons passing through it, regardless of direction; but only the photons in a spatial mode supported by the resonator will pass more than once through the medium and receive substantial amplification.

The light emitted

The light generated by stimulated emission is very similar to the input signal in terms of wavelength, phase, and polarization. This gives laser light its characteristic coherence, and allows it to maintain the uniform polarization and often monochromaticity established by the optical cavity design.

The beam in the cavity and the output beam of the laser, when travelling in free space (or a homogenous medium) rather than waveguides (as in an optical fiber laser), can be approximated as a Gaussian beam in most lasers; such beams exhibit the minimum divergence for a given diameter. However some high power lasers may be multimode, with the transverse modes often approximated using Hermite-Gaussian or Laguerre-Gaussian functions. It has been shown that unstable laser resonators (not used in most lasers) produce fractal shaped beams.[10] Near the beam "waist" (or focal region) it is highly *collimated*: the wavefronts are planar, normal to the direction of propagation, with no beam divergence at that point. However due to diffraction, that can only remain true well within the Rayleigh range. The beam of a single transverse mode (gaussian beam) laser eventually diverges at an angle which varies inversely with the beam diameter, as required by diffraction theory. Thus, the "pencil beam" directly generated by a common helium-neon laser would spread out to a size of perhaps 500 kilometers when shone on the Moon (from the distance of the earth). On the other hand the light from a semiconductor laser typically exits the tiny crystal with a large divergence: up to 50°. However even such a divergent beam can be transformed into a similarly collimated beam by means of a lens system, as is always included, for instance, in a laser pointer whose light originates from a laser diode. That is possible due to the light being of a single spatial mode. This unique property of laser light, spatial coherence, cannot be replicated using standard light sources (except by discarding most of the light) as can be appreciated by comparing the beam from a flashlight (torch) or spotlight to that of almost any laser.

Quantum vs. classical emission processes

The mechanism of producing radiation in a laser relies on stimulated emission, where energy is extracted from a transition in an atom or molecule. This is a quantum phenomenon discovered by Einstein who derived the relationship between the A coefficient describing spontaneous emission and the B coefficient which applies to absorption and stimulated emission. However in the case of the free electron laser, atomic energy levels are not involved; it appears that the operation of this rather exotic device can be explained without reference to quantum mechanics.

Continuous and pulsed modes of operation

A laser can be classified as operating in either continuous or pulsed mode, depending on whether the power output is essentially continuous over time or whether its output takes the form of pulses of light on one or another time scale. Of course even a laser whose output is normally continuous can be intentionally turned on and off at some rate in order to create pulses of light. When the modulation rate is on time scales much slower than the cavity lifetime and the time period over which energy can be stored in the lasing medium or pumping mechanism, then it is still classified as a "modulated" or "pulsed" continuous wave laser. Most laser diodes used in communication systems fall in that category.

Continuous wave operation

Some applications of lasers depend on a beam whose output power is constant over time. Such a laser is known as *continuous wave* (*CW*). Many types of lasers can be made to operate in continuous wave mode to satisfy such an application. Many of these lasers actually lase in several longitudinal modes at the same time, and beats between the slightly different optical frequencies of those oscillations will in fact produce amplitude variations on time scales shorter than the round-trip time (the reciprocal of the frequency spacing between modes), typically a few nanoseconds or less. In most cases these lasers are still termed "continuous wave" as their output power is steady when averaged over any longer time periods, with the very high frequency power variations having little or no impact in the intended application. (However the term is not applied to mode-locked lasers, where the *intention* is to create very short pulses at the rate of the round-trip time).

For continuous wave operation it is required for the population inversion of the gain medium to be continually replenished by a steady pump source. In some lasing media this is impossible. In some other lasers it would require pumping the laser at a very high continuous power level which would be impractical or destroy the laser by producing excessive heat. Such lasers cannot be run in CW mode.

Pulsed operation

Pulsed operation of lasers refers to any laser not classified as continuous wave, so that the optical power appears in pulses of some duration at some repetition rate. This encompasses a wide range of technologies addressing a number of different motivations. Some lasers are pulsed simply because they cannot be run in continuous mode.

In other cases the application requires the production of pulses having as large an energy as possible. Since the pulse energy is equal to the average power divided by the repetition rate, this goal can sometimes be satisfied by lowering the rate of pulses so that more energy can be built up in between pulses. In laser ablation for example, a small volume of material at the surface of a work piece can be evaporated if it is heated in a very short time, whereas supplying the energy gradually would allow for the heat to be absorbed into the bulk of the piece, never attaining a sufficiently high temperature at a particular point.

Other applications rely on the peak pulse power (rather than the energy in the pulse), especially in order to obtain nonlinear optical effects. For a given pulse energy, this requires creating pulses of the shortest possible duration utilizing techniques such as Q-switching.

The optical bandwidth of a pulse cannot be narrower than the reciprocal of the pulse width. In the case of extremely short pulses, that implies lasing over a considerable bandwidth, quite contrary to the very narrow bandwidths typical of CW lasers. The lasing medium in some *dye lasers* and *vibronic solid-state lasers* produces optical gain over a wide bandwidth, making a laser possible which can thus generate pulses of light as short as a few femtoseconds (10^{-15} s).

Q-switching

In a Q-switched laser, the population inversion is allowed to build up by introducing loss inside the resonator which exceeds the gain of the medium; this can also be described as a reduction of the quality factor or 'Q' of the cavity. Then, after the pump energy stored in the laser medium has approached the maximum possible level, the introduced loss mechanism (often an electro- or acousto-optical element) is rapidly removed (or that occurs by itself in a passive device), allowing lasing to begin which rapidly obtains the stored energy in the gain medium. This results in a short pulse incorporating that energy, and thus a high peak power.

Mode-locking

A mode-locked laser is capable of emitting extremely short pulses on the order of tens of picoseconds down to less than 10 femtoseconds. These pulses will repeat at the round trip time, that is, the time that it takes light to complete one round trip between the mirrors comprising the resonator. Due to the Fourier limit (also known as energy-time uncertainty), a pulse of such short temporal length has a spectrum spread over a considerable bandwidth. Thus such a gain medium must have a gain bandwidth sufficiently broad to amplify those frequencies. An example of a suitable material is titanium-doped, artificially grown sapphire (Ti:sapphire) which has a very wide gain bandwidth and can thus produce pulses of only a few femtoseconds duration.

Such mode-locked lasers are a most versatile tool for researching processes occurring on extremely short time scales (known as femtosecond physics, femtosecond chemistry and ultrafast science), for maximizing the effect of nonlinearity in optical materials (e.g. in second-harmonic generation, parametric down-conversion, optical parametric oscillators and the like) due to the large peak power, and in ablation applications. Again, because of the extremely short pulse duration, such a laser will produce pulses which achieve an extremely high peak power.

Pulsed pumping

Another method of achieving pulsed laser operation is to pump the laser material with a source that is itself pulsed, either through electronic charging in the case of flash lamps, or another laser which is already pulsed. Pulsed pumping was historically used with dye lasers where the inverted population lifetime of a dye molecule was so short that a high energy, fast pump was needed. The way to overcome this problem was to charge up large capacitors which are then switched to discharge through flashlamps, producing an intense flash. Pulsed pumping is also required for three-level lasers in which the lower energy level rapidly becomes highly populated preventing further lasing until those atoms relax to the ground state. These lasers, such as the excimer laser and the copper vapor laser, can never be operated in CW mode.

History

Foundations

In 1917, Albert Einstein established the theoretic foundations for the laser and the maser in the paper *Zur Quantentheorie der Strahlung* (On the Quantum Theory of Radiation); via a re-derivation of Max Planck's law of radiation, conceptually based upon probability coefficients (Einstein coefficients) for the absorption, spontaneous emission, and stimulated emission of electromagnetic radiation; in 1928, Rudolf W. Ladenburg confirmed the existences of the phenomena of stimulated emission and negative absorption;[11] in 1939, Valentin A. Fabrikant predicted the use of stimulated emission to amplify “short” waves;[12] in 1947, Willis E. Lamb and R. C. Retherford found apparent stimulated emission in hydrogen spectra and effected the first demonstration of stimulated emission;[11] in 1950, Alfred Kastler (Nobel Prize for Physics 1966) proposed the method of optical pumping, experimentally confirmed, two years later, by Brossel, Kastler, and Winter.[13]

Maser

In 1953, Charles Hard Townes and graduate students James P. Gordon and Herbert J. Zeiger produced the first microwave amplifier, a device operating on similar principles to the laser, but amplifying microwave radiation rather than infrared or visible radiation. Townes's maser was incapable of continuous output. Meanwhile, in the Soviet Union, Nikolay Basov and Aleksandr Prokhorov were independently working on the quantum oscillator and solved the problem of continuous-output systems by using more than two energy levels. These gain media could release stimulated emissions between an excited state and a lower excited state, not the ground state, facilitating the maintenance of a population inversion. In 1955, Prokhorov and Basov suggested optical pumping of a multi-level system as a method for obtaining the population inversion, later a main method of laser pumping.

Aleksandr Prokhorov

Townes reports that several eminent physicists – among them Niels Bohr, John von Neumann, Isidor Rabi, Polykarp Kusch, and Llewellyn Thomas — argued the maser violated Heisenberg's uncertainty principle and hence could not work.[14] In 1964 Charles H. Townes, Nikolay Basov, and Aleksandr Prokhorov shared the Nobel Prize in Physics, "for fundamental work in the field of quantum electronics, which has led to the construction of oscillators and amplifiers based on the maser–laser principle".

Laser

In 1957, Charles Hard Townes and Arthur Leonard Schawlow, then at Bell Labs, began a serious study of the infrared laser. As ideas developed, they abandoned infrared radiation to instead concentrate upon visible light. The concept originally was called an "optical maser". In 1958, Bell Labs filed a patent application for their proposed optical maser; and Schawlow and Townes submitted a manuscript of their theoretical calculations to the *Physical Review*, published that year in Volume 112, Issue No. 6.

Simultaneously, at Columbia University, graduate student Gordon Gould was working on a doctoral thesis about the energy levels of excited thallium. When Gould and Townes met, they spoke of radiation emission, as a general subject; afterwards, in November 1957, Gould noted his ideas for a "laser", including using an open resonator (later an essential laser-device component). Moreover, in 1958, Prokhorov independently proposed using an open resonator, the first published appearance (the USSR) of this idea. Elsewhere, in the U.S., Schawlow and Townes had agreed to an open-resonator laser design – apparently unaware of Prokhorov's publications and Gould's unpublished laser work.

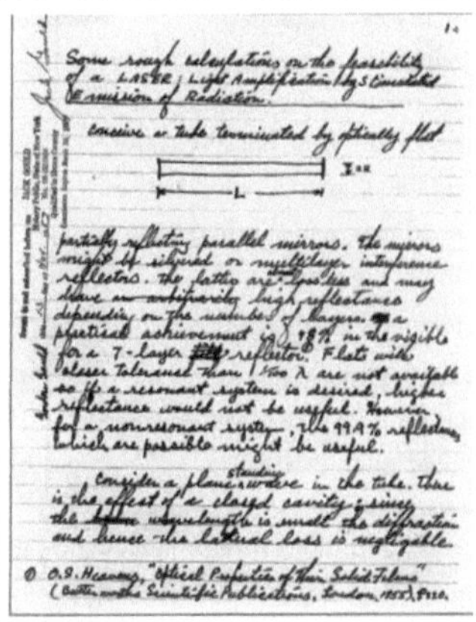

LASER notebook: First page of the notebook wherein Gordon Gould coined the LASER acronym, and described the technologic elements for constructing the device.

At a conference in 1959, Gordon Gould published the term LASER in the paper *The LASER, Light Amplification by Stimulated Emission of Radiation*.[1] [5] Gould's linguistic intention was using the "-aser" word particle as a suffix – to accurately denote the spectrum of the light emitted by the LASER device; thus x-rays: *xaser*, ultraviolet: *uvaser*, et cetera; none established itself as a discrete term, although "raser" was briefly popular for denoting radio-frequency-emitting devices.

Gould's notes included possible applications for a laser, such as spectrometry, interferometry, radar, and nuclear fusion. He continued developing the idea, and filed a patent application in April 1959. The U.S. Patent Office denied his application, and awarded a patent to Bell Labs, in 1960. That provoked a twenty-eight-year lawsuit, featuring scientific prestige and money as the stakes. Gould won his first minor patent in 1977, yet it was not until 1987 that he won the first significant patent lawsuit victory, when a Federal judge ordered the U.S. Patent Office to issue patents to Gould for the optically pumped and the gas discharge laser devices. The question of just how to assign credit for inventing the laser remains unresolved by historians.[15]

On May 16, 1960, Theodore H. Maiman operated the first functioning laser,[16] [17] at Hughes Research Laboratories, Malibu, California, ahead of several research teams, including those of Townes, at Columbia University, Arthur Schawlow, at Bell Labs,[18] and Gould, at the TRG (Technical Research Group) company. Maiman's functional laser used a solid-state flashlamp-pumped synthetic ruby crystal to produce red laser light, at 694 nanometres wavelength; however, the device only was capable of pulsed operation, because of its three-level pumping design scheme. Later in 1960, the Iranian physicist Ali Javan, and William R. Bennett, and Donald Herriott, constructed the first gas laser, using helium and neon that was capable of continuous operation in the infrared (U.S. Patent 3,149,290); later, Javan received the Albert Einstein Award in 1993. Basov and Javan proposed the semiconductor laser diode concept. In 1962, Robert N. Hall demonstrated the first *laser diode* device, made of gallium arsenide and emitted at 850 nm the near-infrared band of the spectrum. Later, in 1962, Nick Holonyak, Jr. demonstrated the first semiconductor laser with a visible emission. This first semiconductor laser could only be used in pulsed-beam operation, and when cooled to liquid nitrogen temperatures (77 K). In 1970, Zhores Alferov, in the USSR, and Izuo Hayashi and Morton Panish of Bell Telephone Laboratories also independently developed room-temperature, continual-operation diode lasers, using the heterojunction structure.

Recent innovations

Since the early period of laser history, laser research has produced a variety of improved and specialized laser types, optimized for different performance goals, including:

- new wavelength bands
- maximum average output power
- maximum peak pulse energy
- maximum peak pulse power
- minimum output pulse duration
- maximum power efficiency
- minimum cost

and this research continues to this day.

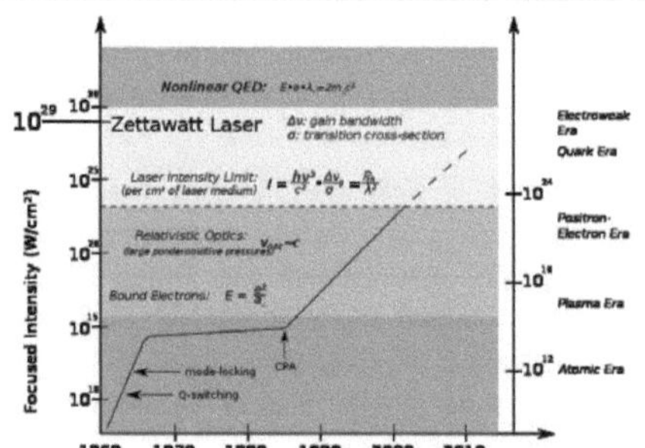

Graph showing the history of maximum laser pulse intensity throughout the past 40 years.

Lasing without maintaining the medium excited into a population inversion was discovered in 1992 in sodium gas and again in 1995 in rubidium gas by various international teams. This was accomplished by using an external maser to induce "optical transparency" in the medium by introducing and destructively interfering the ground electron transitions between two paths, so that the likelihood for the ground electrons to absorb any energy has been cancelled.

Types and operating principles

For a more complete list of laser types see this list of laser types.

Gas lasers

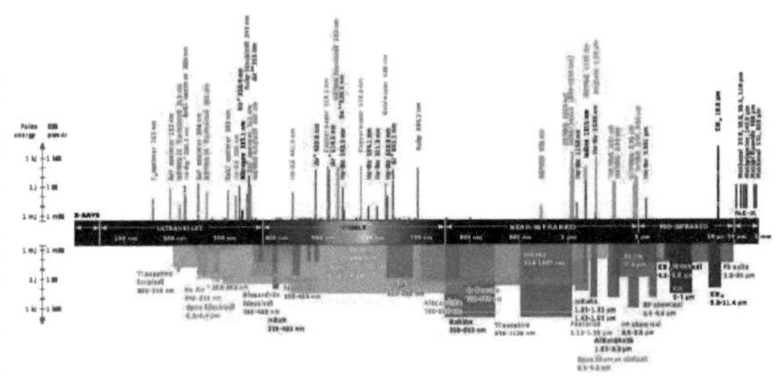

Wavelengths of commercially available lasers. Laser types with distinct laser lines are shown above the wavelength bar, while below are shown lasers that can emit in a wavelength range. The color codifies the type of laser material (see the figure description for more details).

Following the invention of the HeNe gas laser, many other gas discharges have been found to amplify light coherently. Gas lasers using many different gases have been built and used for many purposes. The helium-neon laser (HeNe) is able to operate at a number of different wavelengths, however the vast majority are engineered to lase at 633 nm; these relatively low cost but highly coherent lasers are extremely common in optical research and educational laboratories. Commercial carbon dioxide (CO_2) lasers can emit many hundreds of watts in a single spatial mode which can be concentrated into a tiny spot. This emission is in the thermal infrared at 10.6 μm; such lasers are regularly used in industry for cutting and welding. The efficiency of a CO_2 laser is unusually high: over 10%. Argon-ion lasers can operate at a number of lasing transitions between 351 and 528.7 nm. Depending on the optical design one or more of these transitions can be lasing simultaneously; the most commonly used lines are 458 nm, 488 nm and 514.5 nm. A nitrogen transverse electrical discharge in gas at atmospheric pressure (TEA) laser is an inexpensive gas laser, often home-built by hobbyists, which produces rather incoherent UV light at 337.1 nm.[19] Metal ion lasers are gas lasers that generate deep ultraviolet wavelengths. Helium-silver (HeAg) 224 nm and neon-copper (NeCu) 248 nm are two examples. Like all low-pressure gas lasers, the gain media of these lasers have quite narrow oscillation linewidths, less than 3 GHz (0.5 picometers),[20] making them candidates for use in fluorescence suppressed Raman spectroscopy.

Chemical lasers

Chemical lasers are powered by a chemical reaction permitting a large amount of energy to be released quickly. Such very high power lasers are especially of interest to the military, however continuous wave chemical lasers at very high power levels, fed by streams of gasses, have been developed and have some industrial applications. As examples, in the Hydrogen fluoride laser (2700–2900 nm) and the Deuterium fluoride laser (3800 nm) the reaction is the combination of hydrogen or deuterium gas with combustion products of ethylene in nitrogen trifluoride.

Excimer lasers

Excimer lasers are a special sort of gas laser powered by an electric discharge in which the lasing medium is an excimer, or more precisely an exciplex in existing designs. These are molecules which can only exist with one atom in an excited electronic state. Once the molecule transfers its excitation energy to a photon, therefore, its atoms are no longer bound to each other and the molecule disintegrates. This drastically reduces the population of the lower energy state thus greatly facilitating a population inversion. Excimers currently used are all noble gas compounds; noble gasses are chemically inert and can only form compounds while in an excited state. Excimer lasers typically operate at ultraviolet wavelengths with major applicatons including semiconductor photolithography and LASIK eye surgery. Commonly used excimer molecules include ArF (emission at 193 nm), KrCl (222 nm), KrF (248 nm), XeCl (308 nm), and XeF (351 nm).[21] The molecular fluorine laser, emitting at 157 nm in the vacuum ultraviolet is sometimes referred to as an excimer laser, however this appears to be a misnomer inasmuch as F_2 is a stable compound.

Solid-state lasers

Solid-state lasers use a crystalline or glass rod which is "doped" with ions that provide the required energy states. For example, the first working laser was a ruby laser, made from ruby (chromium-doped corundum). The population inversion is actually maintained in the "dopant", such as chromium or neodymium. These materials are pumped optically using a shorter wavelength than the lasing wavelength, often from a flashtube or from another laser.

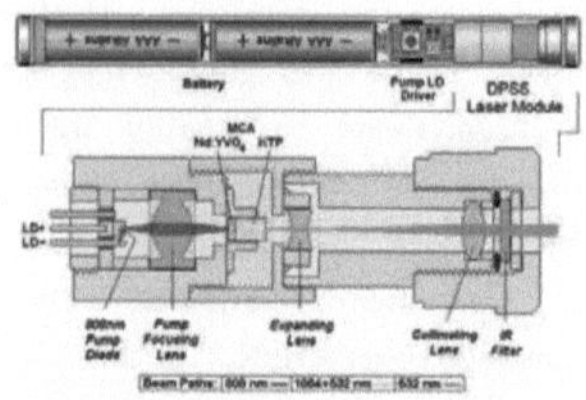

A frequency-doubled green laser pointer, showing internal construction. Two AAA cells and electronics power the laser module (lower diagram) This contains a powerful 808 nm IR diode laser that optically pumps a Nd:YVO_4 crystal inside a laser cavity. That laser produces 1064 nm (infrared) light which is mainly confined inside the resonator. Also inside the laser cavity, however, is a non-linear KTP crystal which causes frequency doubling, resulting in green light at 532 nm. The front mirror is transparent to this visible wavelength which is then expanded and collimated using two lenses (in this particular design).

It should be noted that "solid-state" in this sense refers to a crystal or glass, but this usage is distinct from the designation of "solid-state electronics" in referring to semiconductors. Semiconductor lasers (laser diodes) are pumped electrically and are thus *not* referred to as solid-state lasers. The class of solid-state lasers would, however, properly include fiber lasers in which dopants in the glass lase under optical pumping. But in practice these are simply referred to as "fiber lasers" with "solid-state" reserved for lasers using a solid rod of such a material.

Neodymium is a common "dopant" in various solid-state laser crystals, including yttrium orthovanadate (Nd:YVO_4), yttrium lithium fluoride (Nd:YLF) and yttrium aluminium garnet (Nd:YAG). All these lasers can produce high powers in the infrared spectrum at 1064 nm. They are used for cutting, welding and marking of metals and other materials, and also in spectroscopy and for pumping dye lasers.

These lasers are also commonly frequency doubled, tripled or quadrupled, in so-called "diode pumped solid state" or DPSS lasers. Under second, third, or fourth harmonic generation these produce 532 nm (green, visible), 355 nm and 266 nm (UV) beams. This is the technology behind the bright laser pointers particularly at green (532 nm) and other short visible wavelengths.

Ytterbium, holmium, thulium, and erbium are other common "dopants" in solid-state lasers. Ytterbium is used in crystals such as Yb:YAG, Yb:KGW, Yb:KYW, Yb:SYS, Yb:BOYS, Yb:CaF2, typically operating around 1020–1050 nm. They are potentially very efficient and high powered due to a small quantum defect. Extremely high powers in ultrashort pulses can be achieved with Yb:YAG. Holmium-doped YAG crystals emit at 2097 nm and form an efficient laser operating at infrared wavelengths strongly absorbed by water-bearing tissues. The Ho-YAG is usually operated in a pulsed mode, and passed through optical fiber surgical devices to resurface joints, remove rot from teeth, vaporize cancers, and pulverize kidney and gall stones.

Titanium-doped sapphire (Ti:sapphire) produces a highly tunable infrared laser, commonly used for spectroscopy. It is also notable for use as a mode-locked laser producing ultrashort pulses of extremely high peak power.

Thermal limitations in solid-state lasers arise from unconverted pump power that manifests itself as heat. This heat, when coupled with a high thermo-optic coefficient (dn/dT) can give rise to thermal lensing as well as reduced quantum efficiency. These types of issues can be overcome by another novel diode-pumped solid-state laser, the diode-pumped thin disk laser. The thermal limitations in this laser type are mitigated by using a laser medium geometry in which the thickness is much smaller than the diameter of the pump beam. This allows for a more even thermal gradient in the material. Thin disk lasers have been shown to produce up to kilowatt levels of power.[22]

Fiber lasers

Solid-state lasers or laser amplifiers where the light is guided due to the total internal reflection in a single mode optical fiber are instead called fiber lasers. Guiding of light allows extremely long gain regions providing good cooling conditions; fibers have high surface area to volume ratio which allows efficient cooling. In addition, the fiber's waveguiding properties tend to reduce thermal distortion of the beam. Erbium and ytterbium ions are common active species in such lasers.

Quite often, the fiber laser is designed as a double-clad fiber. This type of fiber consists of a fiber core, an inner cladding and an outer cladding. The index of the three concentric layers is chosen so that the fiber core acts as a single-mode fiber for the laser emission while the outer cladding acts as a highly multimode core for the pump laser. This lets the pump propagate a large amount of power into and through the active inner core region, while still having a high numerical aperture (NA) to have easy launching conditions.

Pump light can be used more efficiently by creating a fiber disk laser, or a stack of such lasers.

Fiber lasers have a fundamental limit in that the intensity of the light in the fiber cannot be so high that optical nonlinearities induced by the local electric field strength can become dominant and prevent laser operation and/or lead to the material destruction of the fiber. This effect is called photodarkening. In bulk laser materials, the cooling is not so efficient, and it is difficult to separate the effects of photodarkening from the thermal effects, but the experiments in fibers show that the photodarkening can be attributed to the formation of long-living color centers.

Photonic crystal lasers

Photonic crystal lasers are lasers based on nano-structures that provide the mode confinement and the density of optical states (DOS) structure required for the feedback to take place. They are typical micrometre-sized and tunable on the bands of the photonic crystals.[23]

Semiconductor lasers

Semiconductor lasers are diodes which are electrically pumped. Recombination of electrons and holes created by the applied current introduces optical gain. Reflection from the ends of the crystal form an optical resonator, although the resonator can be external to the semiconductor in some designs.

A 5.6 mm 'closed can' commercial laser diode, probably from a CD or DVD player

Commercial laser diodes emit at wavelengths from 375 nm to 3500 nm. Low to medium power laser diodes are used in laser printers and CD/DVD players. Laser diodes are also frequently used to optically pump other lasers with high efficiency. The highest power industrial laser diodes, with power up to 10 kW (70dBm), are used in industry for cutting and welding. External-cavity semiconductor lasers have a semiconductor active medium in a larger cavity. These devices can generate high power outputs with good beam quality, wavelength-tunable narrow-linewidth radiation, or ultrashort laser pulses.

Vertical cavity surface-emitting lasers (VCSELs) are semiconductor lasers whose emission direction is perpendicular to the surface of the wafer. VCSEL devices typically have a more circular output beam than conventional laser diodes, and potentially could be much cheaper to manufacture. As of 2005, only 850 nm VCSELs are widely available, with 1300 nm VCSELs beginning to be commercialized,[24] and 1550 nm devices an area of research. VECSELs are external-cavity VCSELs. Quantum cascade lasers are semiconductor lasers that have an active transition between energy *sub-bands* of an electron in a structure containing several quantum wells.

The development of a silicon laser is important in the field of optical computing. Silicon is the material of choice for integrated circuits, and so electronic and silicon photonic components (such as optical interconnects) could be fabricated on the same chip. Unfortunately, silicon is a difficult lasing material to deal with, since it has certain properties which block lasing. However, recently teams have produced silicon lasers through methods such as fabricating the lasing material from silicon and other semiconductor materials, such as indium(III) phosphide or gallium(III) arsenide, materials which allow coherent light to be produced from silicon. These are called hybrid silicon laser. Another type is a Raman laser, which takes advantage of Raman scattering to produce a laser from materials such as silicon.

Dye lasers

Dye lasers use an organic dye as the gain medium. The wide gain spectrum of available dyes, or mixtures of dyes, allows these lasers to be highly tunable, or to produce very short-duration pulses (on the order of a few femtoseconds). Although these tunable lasers are mainly known in their liquid form, researchers have also demonstrated narrow-linewidth tunable emission in dispersive oscillator configurations incorporating solid-state dye gain media.[25] In their most prevalent form these solid state dye lasers use dye-doped polymers as laser media.

Free electron lasers

Free electron lasers, or FELs, generate coherent, high power radiation, that is widely tunable, currently ranging in wavelength from microwaves, through terahertz radiation and infrared, to the visible spectrum, to soft X-rays. They have the widest frequency range of any laser type. While FEL beams share the same optical traits as other lasers, such as coherent radiation, FEL operation is quite different. Unlike gas, liquid, or solid-state lasers, which rely on bound atomic or molecular states, FELs use a relativistic electron beam as the lasing medium, hence the term *free electron*.

Bio laser

Living cells can be genetically engineered to produce Green fluorescent protein (GFP). The GFP is used as the laser's "gain medium", where light amplification takes place. The cells are then placed between two tiny mirrors, just 20 millionths of a metre across, which acted as the "laser cavity" in which light could bounce many times through the cell. Upon bathing the cell with blue light, it could be seen to emit directed and intense green laser light.[26] [27]

Exotic laser media

In September 2007, the BBC News reported that there was speculation about the possibility of using positronium annihilation to drive a very powerful gamma ray laser.[28] Dr. David Cassidy of the University of California, Riverside proposed that a single such laser could be used to ignite a nuclear fusion reaction, replacing the banks of hundreds of lasers currently employed in inertial confinement fusion experiments.[28]

Space-based X-ray lasers pumped by a nuclear explosion have also been proposed as antimissile weapons.[29] [30] Such devices would be one-shot weapons.

Uses

When lasers were invented in 1960, they were called "a solution looking for a problem".[31] Since then, they have become ubiquitous, finding utility in thousands of highly varied applications in every section of modern society, including consumer electronics, information technology, science, medicine, industry, law enforcement, entertainment, and the military.

Lasers range in size from microscopic diode lasers (top) with numerous applications, to football field sized neodymium glass lasers (bottom) used for inertial confinement fusion, nuclear weapons research and other high energy density physics experiments.

The first use of lasers in the daily lives of the general population was the supermarket barcode scanner, introduced in 1974. The laserdisc player, introduced in 1978, was the first successful consumer product to include a laser but the compact disc player was the first laser-equipped device to become common, beginning in 1982 followed shortly by laser printers.

Some other uses are:

- Medicine: Bloodless surgery, laser healing, surgical treatment, kidney stone treatment, eye treatment, dentistry
- Industry: Cutting, welding, material heat treatment, marking parts, non-contact measurement of parts
- Military: Marking targets, guiding munitions, missile defence, electro-optical countermeasures (EOCM), alternative to radar, blinding troops.
- Law enforcement: used for latent fingerprint detection in the forensic identification field[32] [33]
- Research: Spectroscopy, laser ablation, laser annealing, laser scattering, laser interferometry, LIDAR, laser capture microdissection, fluorescence microscopy
- Product development/commercial: laser printers, optical discs (e.g. CDs and the like), barcode scanners, thermometers, laser pointers, holograms, bubblegrams.
- Laser lighting displays: Laser light shows
- Cosmetic skin treatments: acne treatment, cellulite and striae reduction, and hair removal.

In 2004, excluding diode lasers, approximately 131,000 lasers were sold with a value of US$2.19 billion.[34] In the same year, approximately 733 million diode lasers, valued at $3.20 billion, were sold.[35]

Examples by power

Different applications need lasers with different output powers. Lasers that produce a continuous beam or a series of short pulses can be compared on the basis of their average power. Lasers that produce pulses can also be characterized based on the *peak* power of each pulse. The peak power of a pulsed laser is many orders of magnitude greater than its average power. The average output power is always less than the power consumed.

Laser application in astronomical adaptive optics imaging

The continuous or average power required for some uses:

Power	Use
1–5 mW	Laser pointers
5 mW	CD-ROM drive
5–10 mW	DVD player or DVD-ROM drive
100 mW	High-speed CD-RW burner
250 mW	Consumer 16x DVD-R burner
400 mW	Burning through a jewel case including disk within 4 seconds[36]
	DVD 24x dual-layer recording.[37]
1 W	Green laser in current Holographic Versatile Disc prototype development
1–20 W	Output of the majority of commercially available solid-state lasers used for micro machining
30–100 W	Typical sealed CO_2 surgical lasers[38]
100–3000 W	Typical sealed CO_2 lasers used in industrial laser cutting
5 kW	Output power achieved by a 1 cm diode laser bar[39]
100 kW	Claimed output of a CO_2 laser being developed by Northrop Grumman for military (weapon) applications

Examples of pulsed systems with high peak power:

- 700 TW (700×10^{12} W) – National Ignition Facility, a 192-beam, 1.8-megajoule laser system adjoining a 10-meter-diameter target chamber.[40]
- 1.3 PW (1.3×10^{15} W) – world's most powerful laser as of 1998, located at the Lawrence Livermore Laboratory[41]

Hobby uses

In recent years, some hobbyists have taken interests in lasers. Lasers used by hobbyists are generally of class IIIa or IIIb, although some have made their own class IV types.[42] However, compared to other hobbyists, laser hobbyists are far less common, due to the cost and potential dangers involved. Due to the cost of lasers, some hobbyists use inexpensive means to obtain lasers, such as salvaging laser diodes from broken DVD players (red), Blu-ray players (violet), or even higher power laser diodes from CD or DVD burners.[43]

Hobbyists also have been taking surplus pulsed lasers from retired military applications and modifying them for pulsed holography. Pulsed Ruby and pulsed YAG lasers have been used.

Safety

Even the first laser was recognized as being potentially dangerous. Theodore Maiman characterized the first laser as having a power of one "Gillette" as it could burn through one Gillette razor blade. Today, it is accepted that even low-power lasers with only a few milliwatts of output power can be hazardous to human eyesight, when the beam from such a laser hits the eye directly or after reflection from a shiny surface. At wavelengths which the cornea and the lens can focus well, the coherence and low divergence of laser light means that it can be focused by the eye into an extremely small spot on the retina, resulting in localized burning and permanent damage in seconds or even less time.

Warning symbol for lasers

Lasers are usually labeled with a safety class number, which identifies how dangerous the laser is:

- Class I/1 is inherently safe, usually because the light is contained in an enclosure, for example in CD players.
- Class II/2 is safe during normal use; the blink reflex of the eye will prevent damage. Usually up to 1 mW power, for example laser pointers.
- Class IIIa/3R lasers are usually up to 5 mW and involve a small risk of eye damage within the time of the blink reflex. Staring into such a beam for several seconds is likely to cause damage to a spot on the retina.
- Class IIIb/3B can cause immediate eye damage upon exposure.
- Class IV/4 lasers can burn skin, and in some cases, even scattered light can cause eye and/or skin damage. Many industrial and scientific lasers are in this class.

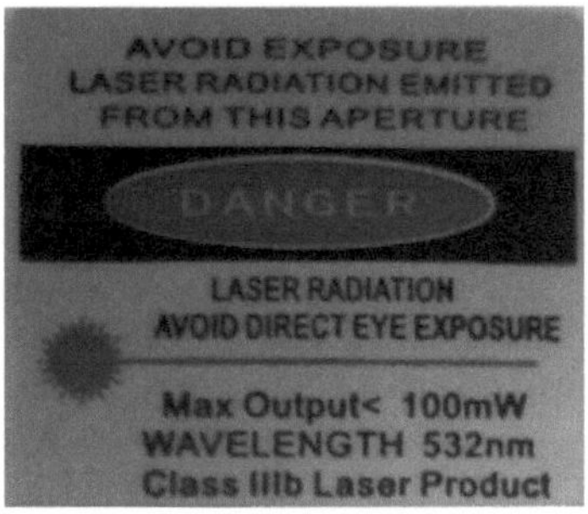

Laser warning label

The indicated powers are for visible-light, continuous-wave lasers. For pulsed lasers and invisible wavelengths, other power limits apply. People working with class 3B and class 4 lasers can protect their eyes with safety goggles which are designed to absorb light of a particular wavelength.

Certain infrared lasers with wavelengths beyond about 1.4 micrometres are often referred to as being "eye-safe". This is because the intrinsic molecular vibrations of water molecules very strongly absorb light in this part of the spectrum, and thus a laser beam at these wavelengths is attenuated so completely as it passes through the eye's cornea that no light remains to be focused by the lens onto the retina. The label "eye-safe" can be misleading, however, as it only applies to relatively low power continuous wave beams; any high power or Q-switched laser at these wavelengths can burn the cornea, causing severe eye damage.

As weapons

Laser beams are famously employed as weapon systems in science fiction, but actual laser weapons are still in the experimental stage. The general idea of laser-beam weaponry is to hit a target with a train of brief pulses of light. The rapid evaporation and expansion of the surface causes shockwaves that damage the target. The power needed to project a high-powered laser beam of this kind is beyond the limit of current mobile power technology thus favoring chemically powered gas dynamic lasers.

Lasers of all but the lowest powers can potentially be used as incapacitating weapons, through their ability to produce temporary or permanent vision loss in varying degrees when aimed at the eyes. The degree, character, and duration of vision impairment caused by eye exposure to laser light varies with the power of the laser, the wavelength(s), the collimation of the beam, the exact orientation of the beam, and the duration of exposure. Lasers of even a fraction of a watt in power can produce immediate, permanent vision loss under certain conditions, making such lasers potential non-lethal but incapacitating weapons. The extreme handicap that laser-induced blindness represents makes the use of lasers even as non-lethal weapons morally controversial, and weapons designed to cause blindness have been banned by the Protocol on Blinding Laser Weapons. The U.S. Air Force is currently working on the Boeing YAL-1 airborne laser, mounted in a Boeing 747, to shoot down enemy ballistic missiles over enemy territory.

In the field of aviation, the hazards of exposure to ground-based lasers deliberately aimed at pilots have grown to the extent that aviation authorities have special procedures to deal with such hazards.[44]

On March 18, 2009 Northrop Grumman claimed that its engineers in Redondo Beach had successfully built and tested an electrically powered solid state laser capable of producing a 100-kilowatt beam, powerful enough to destroy an airplane. According to Brian Strickland, manager for the United States Army's Joint High Power Solid State Laser program, an electrically powered laser is capable of being mounted in an aircraft, ship, or other vehicle because it requires much less space for its supporting equipment than a chemical laser.[45] However the source of such a large electrical power in a mobile application remains unclear.

Fictional predictions

Several novelists described devices similar to lasers, prior to the discovery of stimulated emission:

- A laser-like device was described in Alexey Tolstoy's science fiction novel *The Hyperboloid of Engineer Garin* in 1927.
- Mikhail Bulgakov exaggerated the biological effect (laser bio stimulation) of intensive red light in his science fiction novel *Fatal Eggs* (1925), without any reasonable description of the source of this red light. (In that novel, the red light first appears occasionally from the illuminating system of an advanced microscope; then the protagonist Prof. Persikov arranges the special set-up for generation of the red light.)

See also

- Bessel beam
- Coherent perfect absorber
- dazzler (weapon)
- Free-space optical communication
- Homogeneous broadening
- Induced gamma emission
- Injection seeder
- International Laser Display Association
- Laser accelerometer
- Lasers and aviation safety
- Laser beam profiler
- Laser bonding
- Laser converting
- Laser cooling
- Laser engraving
- Laser medicine
- Laser scalpel
- 3D scanner
- Laser turntable
- Laser beam welding
- List of laser articles
- List of light sources
- Mercury laser
- Nanolaser
- Reference beam
- Rytov number
- Sound Amplification by Stimulated Emission of Radiation SASER
- Selective laser sintering
- Spaser
- Speckle pattern
- Tophat beam

References

Notes

[1] Gould, R. Gordon (1959). "The LASER, Light Amplification by Stimulated Emission of Radiation". In Franken, P.A. and Sands, R.H. (Eds.). *The Ann Arbor Conference on Optical Pumping, the University of Michigan, 15 June through 18 June 1959.* p. 128. OCLC 02460155.

[2] "laser" (http://dictionary.reference.com/browse/laser). Reference.com. . Retrieved May 15, 2008.

[3] Conceptual physics, Paul Hewitt, 2002

[4] "Schawlow and Townes invent the laser" (http://www.bell-labs.com/about/history/laser/). Lucent Technologies. 1998. . Retrieved October 24, 2006.

[5] Chu, Steven; Townes, Charles (2003). "Arthur Schawlow". In Edward P. Lazear (ed.),. *Biographical Memoirs.* vol. 83. National Academy of Sciences. p. 202. ISBN 0-309-08699-X.

[6] "lase" (http://dictionary.reference.com/browse/lase). Dictionary.reference.com. . Retrieved December 10, 2011.

[7] Siegman, Anthony E. (1986). *Lasers.* University Science Books. p. 2. ISBN 0-935702-11-3.

[8] Siegman, Anthony E. (1986). *Lasers.* University Science Books. p. 4. ISBN 0-935702-11-3.

[9] "Nitrogen Laser". *Light and Its Uses.* Scientific American. June 1974. pp. 40–43. ISBN 0-7167-1185-0.

[10] G.P. Karman, G.S. McDonald, G.H.C. New, J.P. Woerdman, " Laser Optics: Fractal modes in unstable resonators (http://www.nature.com/nature/journal/v402/n6758/abs/402138a0.html)", Nature, Vol. 402, 138, November 11, 1999.

[11] Steen, W. M. "Laser Materials Processing", 2nd Ed. 1998.

[12] **(Italian)** "Il rischio da laser: cosa è e come affrontarlo; analisi di un problema non così lontano da noi ("The risk from laser: what it is and what it is like facing it; analysis of a problem which is thus mot far away from us."), Programma Corso di Formazione Obbligatorio anno 2004, Dimitri Batani (Powerpoint presentation >7Mb)" (http://wwwold.unimib.it/ateneo/presentazione/direzione_ammva/prevenzione_protezione/Semin_sicur_laser.ppt). wwwold.unimib.it. . Retrieved January 1, 2007.

[13] The Nobel Prize in Physics 1966 (http://nobelprize.org/nobel_prizes/physics/laureates/1966/press.html) Presentation Speech by Professor Ivar Waller. Retrieved January 1, 2007.

[14] http://books.google.com/books?id=VrbD41GGeJYC&pg=PA69&lpg=PA69&dq=%22niels+bohr%22+rabi+kusch+von+neumann+laser&source=web&ots=0_A7OuramT&sig=4R4yTmk6SmJTN8mZaiOMzgg-LO4

[15] Joan Lisa Bromberg, *The Laser in America, 1950–1970* (1991), pp. 74–77 online (http://www.aip.org/history/exhibits/laser/sections/whoinvented.html)

[16] Maiman, T.H. (1960). "Stimulated optical radiation in ruby". *Nature* **187** (4736): 493–494. Bibcode 1960Natur.187..493M. doi:10.1038/187493a0.

[17] Townes, Charles Hard. "The first laser" (http://www.press.uchicago.edu/Misc/Chicago/284158_townes.html). University of Chicago. . Retrieved May 15, 2008.

[18] Hecht, Jeff (2005). *Beam: The Race to Make the Laser.* Oxford University Press. ISBN 0-19-514210-1.

[19] Csele, Mark (2004). "The TEA Nitrogen Gas Laser" (http://web.archive.org/web/20070911190723/http://www.technology.niagarac.on.ca/people/mcsele/lasers/LasersTEA.htm). *Homebuilt Lasers Page.* Archived from the original (http://www.technology.niagarac.on.ca/people/mcsele/lasers/LasersTEA.htm) on September 11, 2007. . Retrieved September 15, 2007.

[20] "Deep UV Lasers" (http://www.photonsystems.com/pdfs/duv-lasersource.pdf) (PDF). Photon Systems, Covina, Calif. . Retrieved May 27, 2007.

[21] Schuocker, D. (1998). *Handbook of the Eurolaser Academy.* Springer. ISBN 0-412-81910-4.

[22] C. Stewen, M. Larionov, and A. Giesen, “Yb:YAG thin disk laser with 1 kW output power,” in OSA Trends in Optics and Photonics, Advanced Solid-State Lasers, H. Injeyan, U. Keller, and C. Marshall, ed. (Optical Society of America, Washington, DC., 2000) pp. 35–41.

[23] Wu, X.; et al. (October 25, 2004). "Ultraviolet photonic crystal laser" (http://www.eng.yale.edu/images/ArticlPDF/APL04A.PDF). *Applied Physics Letters* **85** (17): 3657. arXiv:physics/0406005. Bibcode 2004ApPhL..85.3657W. doi:10.1063/1.1808888. .

[24] "Picolight ships first 4-Gbit/s 1310-nm VCSEL transceivers" (http://lfw.pennnet.com/Articles/Article_Display.cfm?ARTICLE_ID=243400&p=12), *Laser Focus World*, December 9, 2005. Retrieved May 27, 2006

[25] F. J. Duarte, *Tunable Laser Optics* (Elsevier Academic, New York, 2003) (http://www.opticsjournal.com/tlo.htm).

[26] Palmer, Jason (June 13, 2011). "Laser is produced by a living cell" (http://www.bbc.co.uk/news/science-environment-13725719). *BBC News.* . Retrieved June 13, 2011.

[27] Malte C. Gather & Seok Hyun Yun (June 12, 2011). "Single-cell biological lasers" (http://www.nature.com/nphoton/journal/vaop/ncurrent/full/nphoton.2011.99.html). *Nature Photonics.* . Retrieved June 13, 2011.

[28] Fildes, Jonathan (September 12, 2007). "Mirror particles form new matter" (http://news.bbc.co.uk/2/hi/science/nature/6991030.stm). *BBC News.* . Retrieved May 22, 2008.

[29] Hecht, Jeff (May 2008). "The history of the x-ray laser". *Optics and Photonics News* (Optical Society of America) **19** (5): 26–33.

[30] Robinson, Clarence A. (1981). "Advance made on high-energy laser". *Aviation Week & Space Technology* (February 23, 1981): 25–27.

[31] Charles H. Townes (2003). "The first laser" (http://www.press.uchicago.edu/Misc/Chicago/284158_townes.html). In Laura Garwin and Tim Lincoln. *A Century of Nature: Twenty-One Discoveries that Changed Science and the World.* University of Chicago Press. pp. 107–12. ISBN 0-226-28413-1. . Retrieved February 2, 2008.

[32] Dalrymple BE, Duff JM, Menzel ER. Inherent fingerprint luminescence – detection by laser. Journal of Forensic Sciences, 22(1), 1977, 106–115

[33] Dalrymple BE. Visible and infrared luminescence in documents : excitation by laser. Journal of Forensic Sciences, 28(3), 1983, 692–696
[34] Kincade, Kathy and Stephen Anderson (2005) "Laser Marketplace 2005: Consumer applications boost laser sales 10%", *Laser Focus World*, vol. 41, no. 1. (online (http://lfw.pennnet.com/Articles/Article_Display.cfm?Section=ARCHI&ARTICLE_ID=219847&VERSION_NUM=2&p=12))
[35] Steele, Robert V. (2005) "Diode-laser market grows at a slower rate", *Laser Focus World*, vol. 41, no. 2. (online (http://lfw.pennnet.com/Articles/Article_Display.cfm?Section=ARCHI&ARTICLE_ID=221439&VERSION_NUM=4&p=12))
[36] "Green Laser 400 mW burn a box CD in 4 second" (http://www.youtube.com/watch?v=zhtpYztY8-c). youtube. . Retrieved December 10, 2011.
[37] "Laser Diode Power Output Based on DVD-R/RW specs" (http://elabz.com/laser-diode-power-output-based-on-dvd-rrw-specs/). elabz.com. . Retrieved December 10, 2011.
[38] George M. Peavy, " How to select a surgical veterinary laser (http://www.veterinary-laser.com/expert-opinion.php)", veterinary-laser.com. URL accessed March 14, 2008.
[39] "Cascades™ Horizontal Stacked Arrayes" (http://www.nlight.net/diodes/details/9~Cascades-Horizontal-Stacked-Arrays). nlight.net. . Retrieved March 17, 2011.
[40] Heller, Arnie, " Orchestrating the world's most powerful laser (http://www.llnl.gov/str/JulAug05/VanArsdall.html)." *Science and Technology Review*. Lawrence Livermore National Laboratory, July/August 2005. URL accessed May 27, 2006.
[41] Schewe, Phillip F.; Stein, Ben (November 9, 1998). "Physics News Update 401" (http://newton.ex.ac.uk/aip/physnews.401.html#3). American Institute of Physics. . Retrieved March 15, 2008.
[42] PowerLabs CO_2 LASER! (http://www.powerlabs.org/laser.htm) Sam Barros June 21, 2006. Retrieved January 1, 2007.
[43] "Howto: Make a DVD Burner into a High-Powered Laser" (http://www.felesmagus.com/pages/lasers-howto.html). Felesmagus.com. . Retrieved December 10, 2011.
[44] "Police fight back on laser threat" (http://news.bbc.co.uk/1/hi/technology/7990013.stm). *BBC News*. April 8, 2009. . Retrieved April 4, 2010.
[45] Peter, Pae (March 19, 2009.). "Northrop Advance Brings Era Of The Laser Gun Closer" (http://articles.latimes.com/2009/mar/19/business/fi-laser19). *Los Angeles Times*. p. B2. .

Further reading

Books

- Bertolotti, Mario (1999, trans. 2004). *The History of the Laser*, Institute of Physics. ISBN 0-7503-0911-3
- Csele, Mark (2004). *Fundamentals of Light Sources and Lasers*, Wiley. ISBN 0-471-47660-9
- Koechner, Walter (1992). *Solid-State Laser Engineering*, 3rd ed., Springer-Verlag. ISBN 0-387-53756-2
- Siegman, Anthony E. (1986). *Lasers*, University Science Books. ISBN 0-935702-11-3
- Silfvast, William T. (1996). *Laser Fundamentals*, Cambridge University Press. ISBN 0-521-55617-1
- Svelto, Orazio (1998). *Principles of Lasers*, 4th ed. (trans. David Hanna), Springer. ISBN 0-306-45748-2
- Taylor, Nick (2000). *LASER: The inventor, the Nobel laureate, and the thirty-year patent war*. New York: Simon & Schuster. ISBN 0-684-83515-0.
- Wilson, J. & Hawkes, J.F.B. (1987). *Lasers: Principles and Applications*, Prentice Hall International Series in Optoelectronics, Prentice Hall. ISBN 0-13-523697-5
- Yariv, Amnon (1989). *Quantum Electronics*, 3rd ed., Wiley. ISBN 0-471-60997-8
- Bromberg, Joan Lisa (1991). *The Laser in America, 1950–1970*, MIT Press. ISBN 978-0-262-02318-4

Periodicals

- *Applied Physics B: Lasers and Optics* (ISSN 0946-2171)
- *IEEE Journal of Lightwave Technology* (ISSN 0733-8724)
- *IEEE Journal of Quantum Electronics* (ISSN 0018-9197)
- *IEEE Journal of Selected Topics in Quantum Electronics* (ISSN 1077-260X)
- *IEEE Photonics Technology Letters* (ISSN 1041-1135)
- *Journal of the Optical Society of America B: Optical Physics* (ISSN 0740-3224)
- *Laser Focus World* (ISSN 0740-2511)
- *Optics Letters* (ISSN 0146-9592)
- *Photonics Spectra* (ISSN 0731-1230)

External links

- Encyclopedia of laser physics and technology (http://www.rp-photonics.com/encyclopedia.html) by Dr. Rüdiger Paschotta
- A Practical Guide to Lasers for Experimenters and Hobbyists (http://www.repairfaq.org/sam/lasersam.htm) by Samuel M. Goldwasser
- Homebuilt Lasers Page (http://www.technology.niagarac.on.ca/staff/mcsele/lasers/index.html) by Professor Mark Csele
- Powerful laser is 'brightest light in the universe' (http://space.newscientist.com/article/dn13634-powerful-laser-is-brightest-light-in-the-universe.html?feedId=online-news_rss20) – The world's most powerful laser as of 2008 might create supernova-like shock waves and possibly even antimatter (*New Scientist*, April 9, 2008)
- Homemade laser project (http://www.instructables.com/id/Laser-Flashlight-Hack!!/) by Kip Kedersha
- " The Laser: basic principles (http://prn1.univ-lemans.fr/prn1/siteheberge/optique/M1G1_FBalembois_ang/co/M1G1_anglais_web.html)" an online course by Prof. F. Balembois and Dr. S. Forget. *Instrumentation for Optics*, 2008
- Northrop Grumman's Press Release on the Firestrike 15kw tactical laser product. (http://www.irconnect.com/noc/press/pages/news_releases.html?d=154600)
- Website on Lasers 50th anniversary by APS, OSA, SPIE (http://www.laserfest.org)
- Advancing the Laser anniversary site by SPIE: Video interviews, open-access articles, posters, DVDs (http://www.advancingthelaser.org)
- Bright Idea: The First Lasers (http://www.aip.org/history/exhibits/laser/sections/raydevices.html)
- Free software for Simulation of random laser dynamics (https://nanohub.org/resources/laserdyn)
- Video Demonstrations in Lasers and Optics (http://ocw.mit.edu/resources/res-6-006-video-demonstrations-in-lasers-and-optics-spring-2008/) Produced by the Massachusetts Institute of Technology (MIT). Real-time effects are demonstrated in a way that would be difficult to see in a classroom setting.
- Virtual Museum of Laser History, from the touring exhibit by SPIE (http://spie.org/x39914.xml)

Conformational_change

A macromolecule is usually flexible and dynamic. It can change its shape in response to changes in its environment or other factors; each possible shape is called a conformation, and a transition between them is called a **conformational change**. A macromolecular conformational change may be induced by many factors such as a change in temperature, pH, voltage, ion concentration, phosphorylation, or the binding of a ligand.

Many biophysical techniques such as crystallography, NMR, electron paramagnetic resonance (EPR) using Spin label techniques, Circular Dichroism (CD), Hydrogen Exchange, and FRET can be used to study macromolecular conformational change. Dual Polarisation Interferometry is a benchtop technique capable of measuring conformational changes in biomolecules in real time at very high resolution.

A specific nonlinear optical technique called second-harmonic generation (SHG) has been recently applied to the study of conformational change in proteins. In this method, a second-harmonic-active probe is placed at a site that undergoes motion in the protein by mutagenesis or non-site-specific attachment, and the protein is adsorbed or specifically immobilized to a surface. A change in protein conformation produces a change in the net orientation of the dye relative to the surface plane and therefore the intensity of the second harmonic beam. In a protein sample with a well-defined orientation, the tilt angle of the probe can be quantitatively determined, in real space and real time. Second-harmonic-active unnatural amino acids can also be used as probes.

External links

- Book Chapter on Protein Motions [1] from website of RH Austin [2] at Princeton University
- Frauenfelder, H. New looks at protein motions Nature 338, 623 - 624 (20 April 1989) [3].
- Detection of Protein Conformational Change by Optical Second-Harmonic Generation [4]
- Biodesy [5]

See also

- The Database of Macromolecular Motions (molmovdb)
- Protein dynamics
- Database of protein conformational diversity

References

[1] http://www.physics.princeton.edu/~austin/hf_book/chapter26.pdf
[2] http://www.physics.princeton.edu/~austin/
[3] http://www.nature.com/nature/journal/v338/n6217/abs/338623a0.html
[4] http://scitation.aip.org/vsearch/servlet/VerityServlet?KEY=VIRT02&smode=strresults&CURRENT=&ONLINE=YES&SMODE=&ver=&sti=&possible1=&possible1zone=article&bool1=and&possible2=&possible2zone=multi&bool4=and&possible4=salafsky&possible4zone=author&sort=chron&maxdisp=25&threshold=0&%5Bsearch%5D.x=0&%5Bsearch%5D.y=0&frommonth=&fromday=&fromyear=&tomonth=&today=&toyear=&fromvolume=&fromissue=&tovolume=&toissue=&page=1&origquery=&vdk_query=&chapter=0&docdisp=0
[5] http://www.biodesy.com

Conformation_activity_relationship

Conformation-activity relationship is the relationship between the biological activity and the conformation or conformational changes of a biomolecule. This terminology emphasizes the importance of dynamic conformational changes for the biological function, rather than the importance of static three-dimensional structure used in the analysis of structure activity relationships.[1]

The conformational changes usually take place during intermolecular association, such as protein–protein interaction or protein-ligand binding. A binding partner changes the conformation of a biomolecule (e.g. a protein) to enable or disable its biochemical activity.

Methods for analysis of conformation activity relationship vary from *in silico*[2] or using experimental methods such as X-ray crystallography and NMR where the conformation before and after activity can be compared statically or using dynamic methods such as dual polarisation interferometry or circular dichroism where the kinetics as well as degree of conformational change can be quantified.

Experimental techniques

Static

- X-ray crystallography
- NMR

Dynamic

- Dual polarisation interferometry
- Circular Dichroism

References

[1] Taylor, RE; Chen, Y; Beatty, A; Myles, DC; Zhou, Y (2003). "Conformation-activity relationships in polyketide natural products: a new perspective on the rational design of epothilone analogues". *Journal of the American Chemical Society* **125** (1): 26–7. doi:10.1021/ja028196l. PMID 12515494.

[2] Carotenuto, Alfonso; D'ursi, Anna Maria; Mulinacci, Barbara; Paolini, Ilaria; Lolli, Francesco; Papini, Anna Maria; Novellino, Ettore; Rovero, Paolo (2006). "Conformation–Activity Relationship of Designed Glycopeptides as Synthetic Probes for the Detection of Autoantibodies, Biomarkers of Multiple Sclerosis". *Journal of Medicinal Chemistry* **49** (17): 5072–9. doi:10.1021/jm060117j. PMID 16913697.

Interactome

In molecular biology an **Interactome** is defined as the whole set of molecular interactions in cells. Specifically it means physical interactions among molecules but can also mean indirect interactions among genes, i.e. genetic interactions. It is usually displayed as a directed graph. The word "interactome" was originally coined in 1999 by a group of French scientists headed by Bernard Jacq.[2] The term interactome is partly synonymous with biological network.

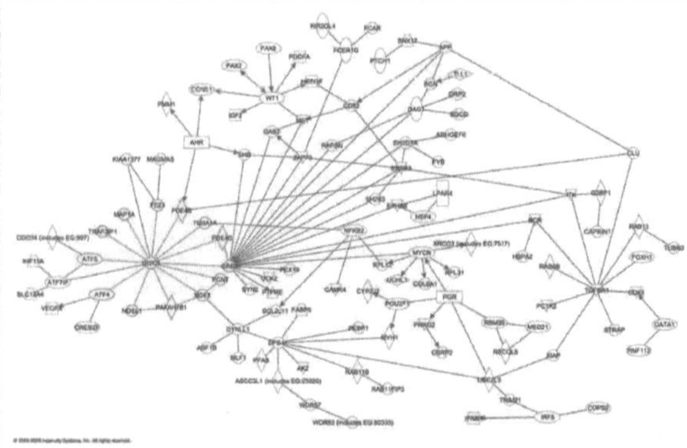

Part of the DISC1 interactome with genes represented by text in boxes and interactions noted by lines between the genes. From Hennah and Porteous, 2009.[1]

Molecular interaction networks

Molecular interactions can occur between molecules belonging to different biochemical families (proteins, nucleic acids, lipids, carbohydrates, etc.) and also within a given family. Whenever such molecules are connected by physical interactions they form molecular interaction networks which are generally classified by the nature of the compounds involved. Most commonly, *interactome* refers to protein–protein interaction (PPI) network (PIN) or subsets thereof. For instance, the Sirt-1 protein interactome [3] is the network involving Sirt-1 and its surrounding proteins. Another extensively studied type of interactome is the protein–DNA interactome, also called a gene-regulatory network, a network formed by transcription factors, chromatin regulatory proteins, and their target genes. Even metabolic networks can be considered as molecular interaction networks: metabolites, i.e. chemical compounds in a cell, are converted into each other by enzymes, which have to bind their substrates physically. In fact, all interactome types are interconnected. For instance, protein interactomes contain many enzymes which in turn form biochemical networks. Similarly, gene regulatory networks overlap substantially with protein interaction networks and signaling networks.

Size of interactomes

-It has been suggested that the size of an organism's interactome correlates better than genome size with the biological complexity of the organism.[4] Although protein–protein interaction maps containing several thousands of binary interactions are now available for several organisms, none of them is presently complete and the size of interactomes is still a matter of debate.

Genetic interaction networks

Genes interact in the sense that they affect each other's function. For instance, a mutation may be harmless, but when it is combined with another mutation, the combination may turn out to be lethal. Such genes are said to "interact genetically". Genes that are connected in such a way form *genetic interaction networks*.

In 2010, the most "complete" gene interactome produced to date was compiled from 54 million two-gene comparisons to describe "the interaction profiles for ~75% of all genes in the Budding yeast," with 170,000 gene interactions.[5]

Although extremely important and useful, the interactome is still being developed and is not complete (as of October 2010). There are various factors that have a role in protein interactions that have yet to be incorporated in the interactome. Many have termed the interactome as a whole as being fuzzy. The binding strength of the various proteins, microenvironmental factors, sensitivity to various procedures, and the physiological state of the cell all affect protein–protein interactions, yet are not accounted for in the interactome. Although the interactome is useful in some ways, it must be analyzed knowing that these factors exist and can affect the protein interactions.[6]

Methods of mapping the interactome

The study of the interactome is called interactomics. The basic unit of a protein network is the protein–protein interaction (PPI). Because the interactome considers the whole cells or organisms, there is a need to collect a massive amount of information.

Experimental methods to identify PPIs: the yeast two hybrid system (Y2H) is suited to explore the binary interactions among two proteins at a time. Affinity purification and subsequent mass spectrometry is suited to identify a protein complex. Both methods can be used in a high-throughput (HTP) fashion.

Predicting PPIs: Using experimental data as a starting point, *homology transfer* is the most straightforward algorithm to predict interactomes. Here, PPIs from one organism are used to predict interactions among homologous proteins in another organism. Other algorithms produce detailed atomic models of protein protein complexes [7] as well as other protein–molecule interactions [8] .

Virus Interactomes

Viral protein interactomes consist of interactions among viral or phage proteins. They were among of the first interactome projects as their genomes are small and all proteins can be analyzed with limited resources. Viral interactomes are connected to their host interactomes, forming virus-host interaction networks [9] . Some published virus interactomes include

- *Escherichia coli* bacteriophage lambda [10]
- *Streptococcus pneumoniae* bacteriophage Dp-1 [11]
- *Streptococcus pneumoniae* bacteriophage Cp-1 [12]
- Human Varicella Zoster Virus (VZV) [13]
- Kaposi's sarcoma-associated herpesvirus (KSHV) [14]
- Epstein-Barr virus (EBV) [15]
- Herpes simplex virus 1 (HSV-1) [16]
- Murine cytomegalovirus (mCMV) [17]

The lambda and VZV interactomes are not only relevant for the biology of these viruses but also for technical reasons: they were the first interactomes that were mapped with multiple Y2H vectors, proving an improved strategy to investigate interactomes more completely than previous attempts have shown.

Bacterial Interactomes

Relatively few bacteria have been comprehensively studied for their protein-protein interactions. Among them are *Treponema pallidum* [18] (the syphilis spirochete) and *Campylobacter jejuni* [19] , which habe been analyzed using extensive Y2H screens. The *Escherichia coli* interactome was studied by large-scale protein complex purification and mass spectrometry [20] .

Eukaryotic Interactomes

There have been several efforts to map eukaryotic interactomes through HTP methods. As of 2006, yeast, fly, worm, and human HTP maps have been created. Recently, pathogen-host interactome (Hepatitis C Virus/Human (2008),[21] Epstein Barr virus/Human (2008), Influenza virus/Human (2009)) was also delineated through HTP to identify essential molecular components for pathoghens but also for the host to recognize pathogens and trigger efficient innate immune response.[22]

Using the interactome

Researchers have begun to use preliminary versions of the interactome to gain understanding about the biology and function of the molecules within them. For example, protein interaction networks have been used to produce improved protein functional annotations (or nannotations) for proteins with unknown functions.[23] [24]

Interactome web servers

- Protinfo PPC [25] predicts the atomic 3D structure of protein protein complexes.[26]

Interactome databases

- APID: Agile Protein Interaction DataAnalyzer [27] – an interactive bioinformatic web-tool that integrates and unifies main known experimentally validated protein–protein interactions (Prieto and De Las Rivas, 2006).
- BioGRID [28] database
- Bioverse [29] database
- ConsensusPathDB includes functional interactions from 12 other databases
- Database of interacting proteins (DIP)[30]
- MIPS [31] database
- PSIMAP [32] database — the first protein structural interactome DB
- InterPare [33] database — a structural protein interfaceome DB
- Biomolecular Interaction Network Database (BIND) [34]
- Online Predicted Human Interaction Database (OPHID) [35]
- Human Protein Reference Database (HPRD [36]) [37]
- HPID: Human Protein Interaction Database [38] (Inha University, Korea)
- MINT: Molecular INTeraction database [39] (Chatr-aryamontri *et al.* 2006)
- PINdb: Proteins Interacting in the Nucleus Database [40] (Luc PV and Tempst P, 2004)
- IntAct: The Molecular Interaction Database [41]
- VirHostNet [42] — **Vir**us-**Host** protein–protein interaction **Net**works knowledgebase
- Interactome Databases [43] — interactome projects at CCSB.
- TRANSFAC, a database about transcription regulating eukaryotic protein-DNA interactions
- PSIbase Database [44] — a global interactome DB based on PSIMAP.
- Interactome.org [45] — a dedicated interactome web site.

See also

- Biological networks
- Bioinformatics
- Omics
- List of omics topics in biology

References

[1] Hennah W, Porteous D (2009). "The DISC1 pathway modulates expression of neurodevelopmental, synaptogenic and sensory perception genes". *PLoS ONE* **4** (3): e4906. doi:10.1371/journal.pone.0004906. PMC 2654149. PMID 19300510.

[2] Sanchez C, Lachaize C, Janody F, *et al.* (January 1999). "Grasping at molecular interactions and genetic networks in Drosophila melanogaster using FlyNets, an Internet database". *Nucleic Acids Res.* **27** (1): 89–94. doi:10.1093/nar/27.1.89. PMC 148104. PMID 9847149.

[3] Sharma, Ankush; Gautam VK, Costantini S, Paladino A and Colonna G (2012 Feb). "Interactomic and pharmacological insights on human Sirt-1" (http://www.frontiersin.org/experimental_pharmacology_and_drug_discovery/10.3389/fphar.2012.00040/abstract). *Front. Pharmacol.* doi:10.3389/fphar.2012.00040. .

[4] Stumpf MP, Thorne T, de Silva E, *et al.* (May 2008). "Estimating the size of the human interactome". *Proc. Natl. Acad. Sci. U.S.A.* **105** (19): 6959–64. doi:10.1073/pnas.0708078105. PMC 2383957. PMID 18474861.

[5] Costanzo M, Baryshnikova A, Bellay J, *et al.* (2010-01-22). "The genetic landscape of a cell". *Science* **327** (5964): 425–431. doi:10.1126/science.1180823. PMID 20093466.

[6] Welch GR (2008). "The Fuzzy Interactome" (http://200.145.134.134/twiki/pub/Main/Miscelanea/0312081.pdf). Cell Press. .

[7] Kittichotirat W, Guerquin M, Bumgarner RE, Samudrala R. (2009.). "Protinfo PPC: A web server for atomic level prediction of protein complexes.". *Nucleic Acids Research* **37** (Web Server issue): W519–W525. doi:10.1093/nar/gkp306. PMC 2703994. PMID 19420059.

[8] McDermott J, Guerquin M, Frazier Z, Chang AN, Samudrala R. (2005). "BIOVERSE: Enhancements to the framework for structural, functional, and contextual annotations of proteins and proteomes.". *Nucleic Acids Research* **33** (Web Server issue): W324–W325. doi:10.1093/nar/gki401. PMC 1160162. PMID 15980482.

[9] Navratil V. et al. (2009). "VirHostNet: a knowledge base for the management and the analysis of proteome-wide virus-host interaction networks". *Nucleic Acids Res.* **37**: D661-8. doi:10.1093/nar/gkn794. PMID 18984613.

[10] Rajagopala SV. et al. (2011). "The protein interaction map of bacteriophage lambda.". *BMC Microbiol.* **11**: 213. doi:10.1186/1471-2180-11-213. PMC 3224144. PMID 21943085.

[11] Sabri M. et al. (2011). "Genome annotation and intraviral interactome for the *Streptococcus pneumoniae* virulent phage Dp-1". *J. Bact.* **193** (2): 551–62. doi:10.1128/JB.01117-10. PMC 3019816. PMID 21097633.

[12] Häuser R. et al. (2011). "The proteome and interactome of *Streptococcus pneumoniae* phage Cp-1". *J. Bact.* **193** (12): 3135–8. doi:10.1128/JB.01481-10. PMC 3133188. PMID 21515781.

[13] Stellberger, T.; et al. (2010). "Improving the yeast two-hybrid system with permutated fusions proteins: the Varicella Zoster Virus interactome". *Proteome Sci. 8:8* **8**: 8. doi:10.1186/1477-5956-8-8. PMC 2832230. PMID 20205919.

[14] Fossum, E; et al. (2009). "Evolutionarily conserved herpesviral protein interaction networks.". *PLoS Pathog. 5(9):e1000570* **5** (9): e1000570. doi:10.1371/journal.ppat.1000570. PMC 2731838. PMID 19730696.

[15] Fossum, E; et al. (2009). "Evolutionarily conserved herpesviral protein interaction networks.". *PLoS Pathog. 5(9):e1000570* **5** (9): e1000570. doi:10.1371/journal.ppat.1000570. PMC 2731838. PMID 19730696.

[16] Fossum, E; et al. (2009). "Evolutionarily conserved herpesviral protein interaction networks.". *PLoS Pathog. 5(9):e1000570* **5** (9): e1000570. doi:10.1371/journal.ppat.1000570. PMC 2731838. PMID 19730696.

[17] Fossum, E; et al. (2009). "Evolutionarily conserved herpesviral protein interaction networks.". *PLoS Pathog. 5(9):e1000570* **5** (9): e1000570. doi:10.1371/journal.ppat.1000570. PMC 2731838. PMID 19730696.

[18] Titz B. et al. (2008). "The binary protein interactome of Treponema pallidum--the syphilis spirochete.". *Plos ONE* **3**: e2292. doi:10.1371/journal.pone.0002292. PMC 2386257. PMID 18509523.

[19] Parrish, JR; et al. (2007). "A proteome-wide protein interaction map for Campylobacter jejuni". *Genome Biol. 8(7): R130* **8** (7): R130. doi:10.1186/gb-2007-8-7-r130. PMC 2323224. PMID 17615063.

[20] Hu, P; et al. (2009). "Global functional atlas of Escherichia coli encompassing previously uncharacterized proteins.". *PLoS Biol. 7(4):e96* **7** (4): e96. doi:10.1371/journal.pbio.1000096. PMC 2672614. PMID 19402753.

[21] de Chassey B, Navratil V, Tafforeau L, *et al.* (2008-11-04). "Hepatitis C virus infection protein network". *Molecular Systems Biology* **4** (4:230): 230. doi:10.1038/msb.2008.66. PMC 2600670. PMID 18985028.

[22] Navratil V, de Chassey B, *et al.* (2010-11-05). "Systems-level comparison of protein–protein interactions between viruses and the human type I interferon system network.". *Journal of Proteome Research* **9** (7): 3527–36. doi:10.1021/pr100326j. PMID 20459142.

[23] Schwikowski, B.; Uetz, P.; Fields, S. (2000). "A network of protein-protein interactions in yeast". *Nature Biotechnology* **18** (12): 1257–1261. doi:10.1038/82360. PMID 11101803.

[24] McDermott J, Bumgarner RE, Samudrala R. (2005). "Functional annotation from predicted protein interaction networks.". *Bioinformatics* **21** (15): 3217–3226. doi:10.1093/bioinformatics/bti514. PMID 15919725.

[25] http://protinfo.compbio.washington.edu/ppc/

[26] Kittichotirat W, Guerquin M, Bumgarner R, Samudrala R. (2009). "Protinfo PPC: A web server for atomic level prediction of protein complexes.". *Nucleic Acids Research* **37**: W519-W525. doi:10.1093/nar/gkp306.

[27] http://bioinfow.dep.usal.es/apid/

[28] http://www.thebiogrid.org/

[29] http://bioverse.compbio.washington.edu/

[30] Xenarios I, Rice DW, Salwinski L, Baron MK, Marcotte EM, Eisenberg D (2000). "DIP: the database of interacting proteins". *Nucleic Acids Res.* **28** (1): 289–91. doi:10.1093/nar/28.1.289. PMC 102387. PMID 10592249.

[31] http://mips.gsf.de/services/ppi

[32] http://psimap.com

[33] http://interpare.net

[34] Bader GD *et al.* (2003). "BIND: The Biomolecular interaction Network Database". *Nucleic Acids Res.* **31** (1): 248–50. doi:10.1093/nar/gkg056. PMC 165503. PMID 12519993.

[35] Brown KR, Jurisica I (2005). "Online Predicted Human Interaction Database". *Bioinformatics* **21** (9): 2076–82. doi:10.1093/bioinformatics/bti273. PMID 15657099.

[36] http://hprd.org:

[37] *Peri S., Navarro JD, Amanchy R, Kristiansen TZ, Jonnalagadda CK *et al.* (2003). "Development of human protein reference database as an initial platform for approaching systems biology in humans". *Genome Res* **13** (10): 2363–71. doi:10.1101/gr.1680803. PMC 403728. PMID 14525934.

[38] http://wilab.inha.ac.kr/hpid/

[39] http://mint.bio.uniroma2.it/mint/Welcome.do

[40] http://pin.mskcc.org

[41] http://www.ebi.ac.uk/intact

[42] http://pbildb1.univ-lyon1.fr/virhostnet

[43] http://interactome.dfci.harvard.edu

[44] http://psimap.com/

[45] http://interactome.org

Further reading

- De Las Rivas J, Fontanillo C (June 2010). "Protein–Protein Interactions Essentials: Key Concepts to Building and Analyzing Interactome Networks". *PLoS Computational Biology* **6** (6): e1000807. doi:10.1371/journal.pcbi.1000807. PMC 2891586. PMID 20589078..
- Park J, Lappe M, Teichmann SA (Mar 2001). "Mapping protein family interactions: intramolecular and intermolecular protein family interaction repertoires in the PDB and yeast". *J Mol Biol.* **307** (3): 929–38. doi:10.1006/jmbi.2001.4526. PMID 11273711.

Biochemistry

Biochemistry, sometimes called **biological chemistry**, is the study of chemical processes in living organisms, including, but not limited to, living matter. The laws of biochemistry govern all living organisms and living processes. By controlling information flow through biochemical signalling and the flow of chemical energy through metabolism, biochemical processes give rise to the complexity of life.

Much of biochemistry deals with the structures, functions and interactions of cellular components such as proteins, carbohydrates, lipids, nucleic acids and other biomolecules —although increasingly processes rather than individual molecules are the main focus. Among the vast number of different biomolecules, many are complex and large molecules (called *biopolymers*), which are composed of similar repeating subunits (called *monomers*). Each class of polymeric biomolecule has a different set of subunit types.[1] For example, a protein is a polymer whose subunits are selected from a set of 20 or more amino acids. Biochemistry studies the chemical properties of important biological molecules, like proteins, and in particular the chemistry of enzyme-catalyzed reactions.

The biochemistry of cell metabolism and the endocrine system has been extensively described. Other areas of biochemistry include the genetic code (DNA, RNA), protein synthesis, cell membrane transport and signal transduction.

Over the last 40 years biochemistry has become so successful at explaining living processes that now almost all areas of the life sciences from botany to medicine are engaged in biochemical research. Today the main focus of pure biochemistry is in understanding how biological molecules give rise to the processes that occur within living cells, which in turn relates greatly to the study and understanding of whole organisms.

History

It once was generally believed that life and its materials had some essential property or substance distinct from any found in non-living matter, and it was thought that only living beings could produce the molecules of life. Then, in 1828, Friedrich Wöhler published a paper on the synthesis of urea, proving that organic compounds can be created artificially.[2] [3]

Gerty Cori and Carl Cori jointly won the Nobel Prize in 1947 for their discovery of the Cori cycle at RPMI.

The dawn of biochemistry may have been the discovery of the first enzyme, diastase (today called amylase), in 1833 by Anselme Payen. Eduard Buchner contributed the first demonstration of a complex biochemical process outside of a cell in 1896: alcoholic fermentation in cell extracts of yeast. Although the term "biochemistry" seems to have been first used in 1882, it is generally accepted that the formal coinage of biochemistry occurred in 1903 by Carl Neuberg, a German chemist. Previous to this time, this area would have been referred to as physiological chemistry. Since then, biochemistry has advanced, especially since the mid-20th century, with the development of new techniques such as chromatography, X-ray diffraction, dual polarisation interferometry, NMR spectroscopy, radioisotopic labeling, electron microscopy, and molecular dynamics simulations. These techniques allowed for the discovery and detailed analysis of many molecules and metabolic pathways of the cell, such as glycolysis and the Krebs cycle (citric acid cycle).

Another significant historic event in biochemistry is the discovery of the gene and its role in the transfer of information in the cell. This part of biochemistry is often called molecular biology. In the 1950s, James D. Watson, Francis Crick, Rosalind Franklin, and Maurice Wilkins were instrumental in solving DNA structure and suggesting its relationship with genetic transfer of information. In 1958, George Beadle and Edward Tatum received the Nobel Prize for work in fungi showing that one gene produces one enzyme. In 1988, Colin Pitchfork was the first person convicted of murder with DNA evidence, which led to growth of forensic science. More recently, Andrew Z. Fire and Craig C. Mello received the 2006 Nobel Prize for discovering the role of RNA interference (RNAi), in the silencing of gene expression.

Starting materials: the chemical elements of life

Around two dozen of the 94 naturally-occurring chemical elements are essential to various kinds of biological life. Most rare elements on Earth are not needed by life (exceptions being selenium and iodine), while a few common ones (aluminum and titanium) are not used. Most organisms share element needs, but there are a few differences between plants and animals. For example ocean algae use bromine but land plants and animals seem to need none. All animals require sodium, but some plants do not. Plants need boron and silicon, but animals may not (or may need ultra-small amounts).

Just six elements—carbon, hydrogen, nitrogen, oxygen, calcium, and phosphorus—make up almost 99% of the mass of a human body (see composition of the human body for a complete list). In addition to the six major elements that compose most of the human body, humans require smaller amounts of possibly 18 more.[4]

Biomolecules

The four main classes of molecules in biochemistry are carbohydrates, lipids, proteins, and nucleic acids. Many biological molecules are polymers: in this terminology, ***monomers*** are relatively small micromolecules that are linked together to create large macromolecules, which are known as ***polymers***. When monomers are linked together to synthesize a biological polymer, they undergo a process called dehydration synthesis. Different macromolecules can assemble in larger complexes, often needed for biological activity.

Carbohydrates

Carbohydrates are made from monomers called *monosaccharides*. Some of these monosaccharides include glucose ($C_6H_{12}O_6$), fructose ($C_6H_{12}O_6$), and deoxyribose ($C_5H_{10}O_4$). When two monosaccharides undergo dehydration synthesis, water is produced, as two hydrogen atoms and one oxygen atom are lost from the two monosaccharides' hydroxyl group.

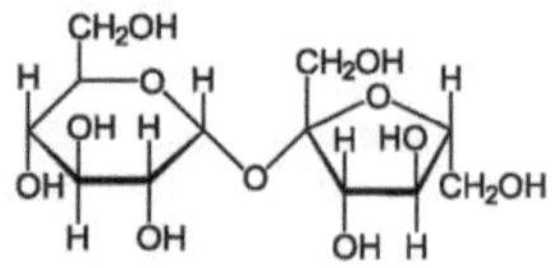

A molecule of sucrose (glucose + fructose), a disaccharide.

Lipids

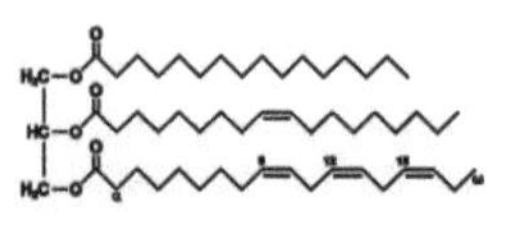

A triglyceride with a glycerol molecule on the left and three fatty acids coming off it.

Lipids are usually made from one molecule of glycerol combined with other molecules. In triglycerides, the main group of bulk lipids, there is one molecule of glycerol and three fatty acids. Fatty acids are considered the monomer in that case, and may be saturated (no double bonds in the carbon chain) or unsaturated (one or more double bonds in the carbon chain).

Lipids, especially phospholipids, are also used in various pharmaceutical products, either as co-solubilisers (e.g., in parenteral infusions) or else as drug carrier components (e.g., in a liposome or transfersome).

Proteins

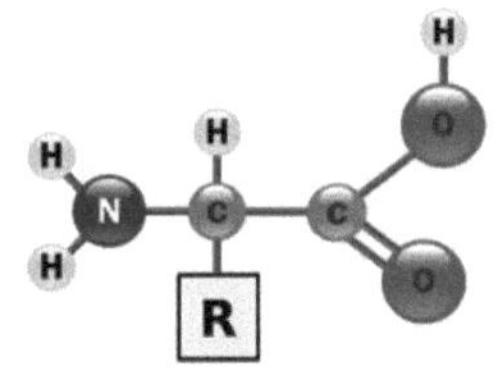

The general structure of an α-amino acid, with the amino group on the left and the carboxyl group on the right.

Proteins are very large molecules – macro-biopolymers – made from monomers called *amino acids*. There are 20 standard amino acids, each containing a carboxyl group, an amino group, and a side-chain (known as an "R" group). The "R" group is what makes each amino acid different, and the properties of the side-chains greatly influence the overall three-dimensional conformation of a protein. When amino acids combine, they form a special bond called a peptide bond through dehydration synthesis, and become a ***polypeptide***, or protein.

In order to determine whether two proteins are related, or in other words to decide whether they are homologous or not, scientists use sequence-comparison methods. Methods like Sequence Alignments and Structural Alignments are powerful tools that help scientists identify homologies between related molecules.

The relevance of finding homologies among proteins goes beyond forming an evolutionary pattern of protein families. By finding how similar two protein sequences are, we acquire knowledge about their structure and therefore their function.

Nucleic acids

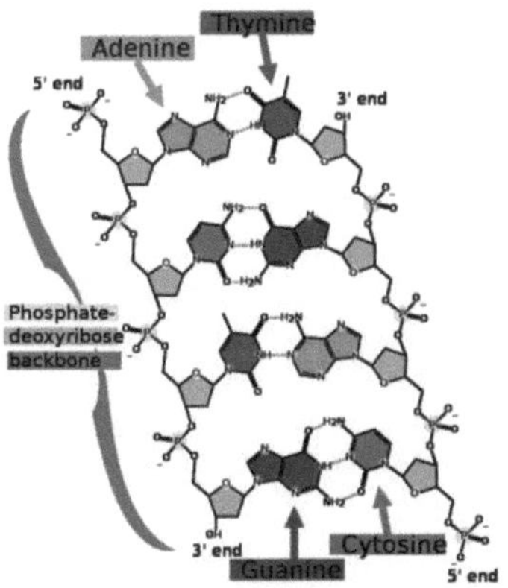

The structure of deoxyribonucleic acid (DNA), the picture shows the monomers being put together.

Nucleic acids are the molecules that make up DNA, an extremely important substance that all cellular organisms use to store their genetic information. The most common nucleic acids are deoxyribonucleic acid and ribonucleic acid. Their monomers are called nucleotides. The most common nucleotides are adenine, cytosine, guanine, thymine, and uracil. Adenine binds with thymine and uracil; Thymine binds only with adenine; and cytosine and guanine can bind only with each other.

Carbohydrates

The function of carbohydrates includes energy storage and providing structure. Sugars are carbohydrates, but not all carbohydrates are sugars. There are more carbohydrates on Earth than any other known type of biomolecule; they are used to store energy and genetic information, as well as play important roles in cell to cell interactions and communications.

Monosaccharides

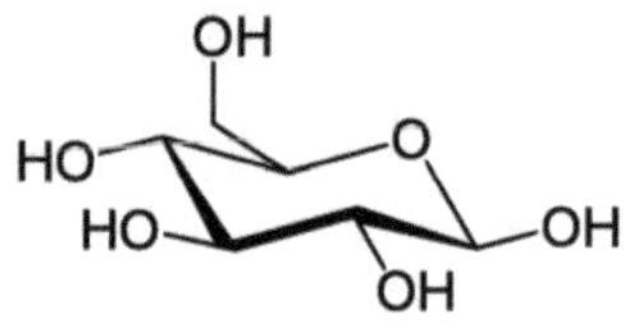

Glucose

The simplest type of carbohydrate is a monosaccharide, which among other properties contains carbon, hydrogen, and oxygen, mostly in a ratio of 1:2:1 (generalized formula $C_nH_{2n}O_n$, where *n* is at least 3). Glucose, one of the most important carbohydrates, is an example of a monosaccharide. So is fructose, the sugar commonly associated with the sweet taste of fruits.[5] [a] Some carbohydrates (especially after condensation to oligo- and polysaccharides) contain less carbon relative to H and O, which still are present in 2:1 (H:O) ratio. Monosaccharides can be grouped into aldoses (having an aldehyde group at the end of the chain, e.g. glucose) and ketoses (having a keto group in their chain; e.g. fructose). Both aldoses and ketoses occur in an equilibrium (starting with chain lengths of C4) cyclic forms. These are generated by bond formation between one of the hydroxyl groups of the sugar chain with the carbon of the aldehyde or keto group to form a hemiacetal bond. This leads to saturated five-membered (in furanoses) or six-membered (in pyranoses) heterocyclic rings containing one O as heteroatom.

Disaccharides

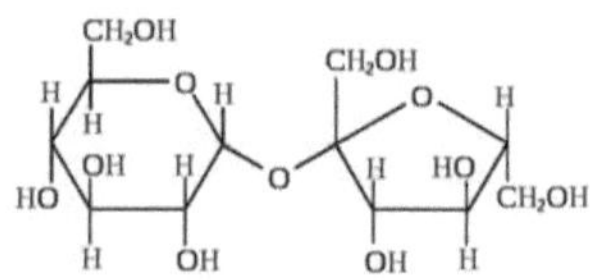

Sucrose: ordinary table sugar and probably the most familiar carbohydrate.

Two monosaccharides can be joined together using dehydration synthesis, in which a hydrogen atom is removed from the end of one molecule and a hydroxyl group (—OH) is removed from the other; the remaining residues are then attached at the sites from which the atoms were removed. The H—OH or H_2O is then released as a molecule of water, hence the term *dehydration*. The new molecule, consisting of two monosaccharides, is called a *disaccharide* and is conjoined together by a glycosidic or ether bond. The reverse reaction can also occur, using a molecule of water to split up a disaccharide and break the glycosidic bond; this is termed *hydrolysis*. The most well-known disaccharide is sucrose, ordinary sugar (in scientific contexts, called *table sugar* or *cane sugar* to differentiate it from other sugars). Sucrose consists of a glucose molecule and a fructose molecule joined together. Another important disaccharide is lactose, consisting of a glucose molecule and a galactose molecule. As most humans age, the production of lactase, the enzyme that hydrolyzes lactose back into glucose and galactose, typically decreases. This results in lactase deficiency, also called *lactose intolerance*.

Sugar polymers are characterised by having reducing or non-reducing ends. A reducing end of a carbohydrate is a carbon atom that can be in equilibrium with the open-chain aldehyde or keto form. If the joining of monomers takes place at such a carbon atom, the free hydroxy group of the pyranose or furanose form is exchanged with an OH-side-chain of another sugar, yielding a full acetal. This prevents opening of the chain to the aldehyde or keto form and renders the modified residue non-reducing. Lactose contains a reducing end at its glucose moiety, whereas the galactose moiety form a full acetal with the C4-OH group of glucose. Saccharose does not have a reducing end because of full acetal formation between the aldehyde carbon of glucose (C1) and the keto carbon of fructose (C2).

Oligosaccharides and polysaccharides

When a few (around three to six) monosaccharides are joined together, it is called an *oligosaccharide* (*oligo-* meaning "few"). These molecules tend to be used as markers and signals, as well as having some other uses. Many monosaccharides joined together make a polysaccharide. They can be joined together in one long linear chain, or they may be branched. Two of the most common polysaccharides are cellulose and glycogen, both consisting of repeating glucose monomers.

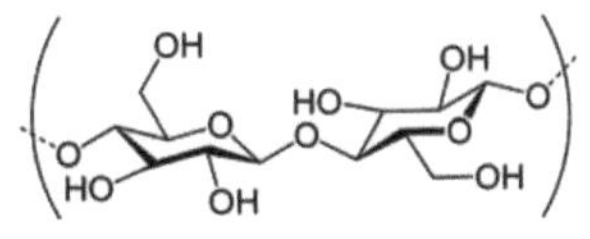

Cellulose as polymer of β-D-glucose

- *Cellulose* is made by plants and is an important structural component of their cell walls. Humans can neither manufacture nor digest it.
- *Glycogen*, on the other hand, is an animal carbohydrate; humans and other animals use it as a form of energy storage.

Use of carbohydrates as an energy source

Glucose is the major energy source in most life forms. For instance, polysaccharides are broken down into their monomers (glycogen phosphorylase removes glucose residues from glycogen). Disaccharides like lactose or sucrose are cleaved into their two component monosaccharides.

Glycolysis (anaerobic)

Glucose is mainly metabolized by a very important ten-step pathway called glycolysis, the net result of which is to break down one molecule of glucose into two molecules of pyruvate; this also produces a net two molecules of ATP, the energy currency of cells, along with two reducing equivalents in the form of converting NAD^+ to NADH. This does not require oxygen; if no oxygen is available (or the cell cannot use oxygen), the NAD is restored by converting the pyruvate to lactate (lactic acid) (e.g., in humans) or to ethanol plus carbon dioxide (e.g., in yeast). Other monosaccharides like galactose and fructose can be converted into intermediates of the glycolytic pathway.

Aerobic

In aerobic cells with sufficient oxygen, as in most human cells, the pyruvate is further metabolized. It is irreversibly converted to acetyl-CoA, giving off one carbon atom as the waste product carbon dioxide, generating another reducing equivalent as NADH. The two molecules acetyl-CoA (from one molecule of glucose) then enter the citric acid cycle, producing two more molecules of ATP, six more NADH molecules and two reduced (ubi)quinones (via $FADH_2$ as enzyme-bound cofactor), and releasing the remaining carbon atoms as carbon dioxide. The produced NADH and quinol molecules then feed into the enzyme complexes of the respiratory chain, an electron transport system transferring the electrons ultimately to oxygen and conserving the released energy in the form of a proton gradient over a membrane (inner mitochondrial membrane in eukaryotes). Thus, oxygen is reduced to water and the original electron acceptors NAD^+ and quinone are regenerated. This is why humans breathe in oxygen and breathe out carbon dioxide. The energy released from transferring the electrons from high-energy states in NADH and quinol is conserved first as proton gradient and converted to ATP via ATP synthase. This generates an additional *28* molecules of ATP (24 from the 8 NADH + 4 from the 2 quinols), totaling to 32 molecules of ATP conserved per degraded glucose (two from glycolysis + two from the citrate cycle). It is clear that using oxygen to completely oxidize glucose provides an organism with far more energy than any oxygen-independent metabolic feature, and this is thought to be the reason why complex life appeared only after Earth's atmosphere accumulated large amounts of oxygen.

Gluconeogenesis

In vertebrates, vigorously contracting skeletal muscles (during weightlifting or sprinting, for example) do not receive enough oxygen to meet the energy demand, and so they shift to anaerobic metabolism, converting glucose to lactate. The liver regenerates the glucose, using a process called gluconeogenesis. This process is not quite the opposite of glycolysis, and actually requires three times the amount of energy gained from glycolysis (six molecules of ATP are used, compared to the two gained in glycolysis). Analogous to the above reactions, the glucose produced can then undergo glycolysis in tissues that need energy, be stored as glycogen (or starch in plants), or be converted to other monosaccharides or joined into di- or oligosaccharides. The combined pathways of glycolysis during exercise, lactate's crossing via the bloodstream to the liver, subsequent gluconeogenesis and release of glucose into the bloodstream is called the Cori cycle.

Proteins

A schematic of hemoglobin. The red and blue ribbons represent the protein globin; the green structures are the heme groups.

Like carbohydrates, some proteins perform largely structural roles. For instance, movements of the proteins actin and myosin ultimately are responsible for the contraction of skeletal muscle. One property many proteins have is that they specifically bind to a certain molecule or class of molecules—they may be *extremely* selective in what they bind. Antibodies are an example of proteins that attach to one specific type of molecule. In fact, the enzyme-linked immunosorbent assay (ELISA), which uses antibodies, is currently one of the most sensitive tests modern medicine uses to detect various biomolecules. Probably the most important proteins, however, are the enzymes. These molecules recognize specific reactant molecules called *substrates*; they then catalyze the reaction between them. By lowering the activation energy, the enzyme speeds up that reaction by a rate of 10^{11} or more: a reaction that would normally take over 3,000 years to complete spontaneously might take less than a second with an enzyme. The enzyme itself is not used up in the process, and is free to catalyze the same reaction with a new set of substrates. Using various modifiers, the activity of the enzyme can be regulated, enabling control of the biochemistry of the cell as a whole.

In essence, proteins are chains of amino acids. An amino acid consists of a carbon atom bound to four groups. One is an amino group, —NH_2, and one is a carboxylic acid group, —COOH (although these exist as —NH_3^+ and —COO^- under physiologic conditions). The third is a simple hydrogen atom. The fourth is commonly denoted "—R" and is different for each amino acid. There are twenty standard amino acids. Some of these have functions by themselves or in a modified form; for instance, glutamate functions as an important neurotransmitter.

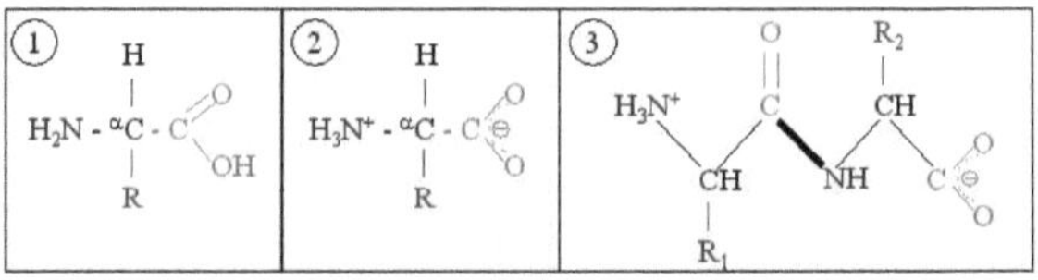

Generic amino acids (1) in neutral form, (2) as they exist physiologically, and (3) joined together as a dipeptide.

Amino acids can be joined together via a peptide bond. In this dehydration synthesis, a water molecule is removed and the peptide bond connects the nitrogen of one amino acid's amino group to the carbon of the other's carboxylic acid group. The resulting molecule is called a *dipeptide*, and short stretches of amino acids (usually, fewer than thirty) are called *peptides* or polypeptides. Longer stretches merit the title *proteins*. As an example, the important blood serum protein albumin contains 585 amino acid residues.

The structure of proteins is traditionally described in a hierarchy of four levels. The primary structure of a protein simply consists of its linear sequence of amino acids; for instance, "alanine-glycine-tryptophan-serine-glutamate-asparagine-glycine-lysine-...". Secondary structure is concerned with local morphology (morphology being the study of structure). Some combinations of amino acids will tend to curl up in a coil called an α-helix or into a sheet called a β-sheet; some α-helixes can be seen in the hemoglobin schematic above. Tertiary structure is the entire three-dimensional shape of the protein. This shape is determined by the sequence of amino acids. In fact, a single change can change the entire structure. The alpha chain of hemoglobin contains 146 amino acid residues; substitution of the glutamate residue at position 6 with a valine residue changes the behavior of hemoglobin so much that it results in sickle-cell disease. Finally, quaternary structure is concerned with the structure of a protein with multiple peptide subunits, like hemoglobin with its four subunits. Not all proteins have more than one subunit.

Ingested proteins are usually broken up into single amino acids or dipeptides in the small intestine, and then absorbed. They can then be joined together to make new proteins. Intermediate products of glycolysis, the citric acid cycle, and the pentose phosphate pathway can be used to make all twenty amino acids, and most bacteria and plants possess all the necessary enzymes to synthesize them. Humans and other mammals, however, can synthesize only half of them. They cannot synthesize isoleucine, leucine, lysine, methionine, phenylalanine, threonine, tryptophan, and valine. These are the essential amino acids, since it is essential to ingest them. Mammals do possess the enzymes to synthesize alanine, asparagine, aspartate, cysteine, glutamate, glutamine, glycine, proline, serine, and tyrosine, the nonessential amino acids. While they can synthesize arginine and histidine, they cannot produce it in sufficient amounts for young, growing animals, and so these are often considered essential amino acids.

If the amino group is removed from an amino acid, it leaves behind a carbon skeleton called an α-keto acid. Enzymes called transaminases can easily transfer the amino group from one amino acid (making it an α-keto acid) to another α-keto acid (making it an amino acid). This is important in the biosynthesis of amino acids, as for many of the pathways, intermediates from other biochemical pathways are converted to the α-keto acid skeleton, and then an amino group is added, often via transamination. The amino acids may then be linked together to make a protein.

A similar process is used to break down proteins. It is first hydrolyzed into its component amino acids. Free ammonia (NH_3), existing as the ammonium ion (NH_4^+) in blood, is toxic to life forms. A suitable method for excreting it must therefore exist. Different strategies have evolved in different animals, depending on the animals' needs. Unicellular organisms, of course, simply release the ammonia into the environment. Likewise, bony fish can release the ammonia into the water where it is quickly diluted. In general, mammals convert the ammonia into urea, via the urea cycle.

Lipids

The term lipid comprises a diverse range of molecules and to some extent is a catchall for relatively water-insoluble or nonpolar compounds of biological origin, including waxes, fatty acids, fatty-acid derived phospholipids, sphingolipids, glycolipids, and terpenoids (e.g., retinoids and steroids). Some lipids are linear aliphatic molecules, while others have ring structures. Some are aromatic, while others are not. Some are flexible, while others are rigid.

Most lipids have some polar character in addition to being largely nonpolar. In general, the bulk of their structure is nonpolar or hydrophobic ("water-fearing"), meaning that it does not interact well with polar solvents like water. Another part of their structure is polar or hydrophilic ("water-loving") and will tend to associate with polar solvents like water. This makes them amphiphilic molecules (having both hydrophobic and hydrophilic portions). In the case of cholesterol, the polar group is a mere -OH (hydroxyl or alcohol). In the case of phospholipids, the polar groups are considerably larger and more polar, as described below.

Lipids are an integral part of our daily diet. Most oils and milk products that we use for cooking and eating like butter, cheese, ghee etc., are composed of fats. Vegetable oils are rich in various polyunsaturated fatty acids (PUFA). Lipid-containing foods undergo digestion within the body and are broken into fatty acids and glycerol, which are the final degradation products of fats and lipids.

Nucleic acids

A nucleic acid is a complex, high-molecular-weight biochemical macromolecule composed of nucleotide chains that convey genetic information. The most common nucleic acids are deoxyribonucleic acid (DNA) and ribonucleic acid (RNA). Nucleic acids are found in all living cells and viruses. Aside from the genetic material of the cell, nucleic acids often play a role as second messengers, as well as forming the base molecule for adenosine triphosphate, the primary energy-carrier molecule found in all living organisms.

Nucleic acid, so called because of its prevalence in cellular nuclei, is the generic name of the family of biopolymers. The monomers are called nucleotides, and each consists of three components: a nitrogenous heterocyclic base (either a purine or a pyrimidine), a pentose sugar, and a phosphate group. Different nucleic acid types differ in the specific sugar found in their chain (e.g., DNA or deoxyribonucleic acid contains 2-deoxyriboses). Also, the nitrogenous bases possible in the two nucleic acids are different: adenine, cytosine, and guanine occur in both RNA and DNA, while thymine occurs only in DNA and uracil occurs in RNA.

Relationship to other "molecular-scale" biological sciences

Researchers in biochemistry use specific techniques native to biochemistry, but increasingly combine these with techniques and ideas developed in the fields of genetics, molecular biology and biophysics. There has never been a hard-line between these disciplines in terms of content and technique. Today, the terms *molecular biology* and *biochemistry* are nearly interchangeable. The following figure is a schematic that depicts one possible view of the relationship between the fields:

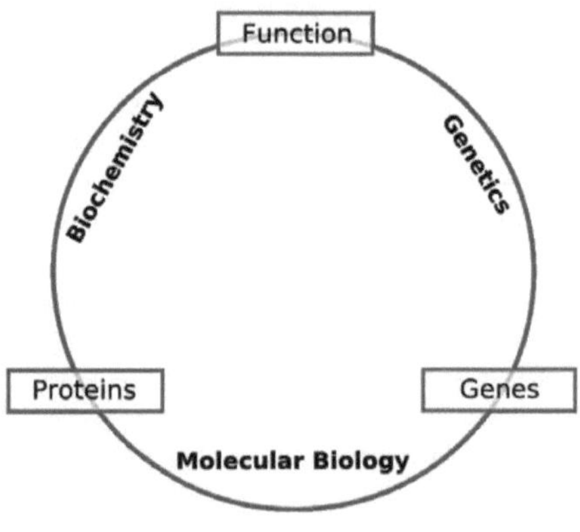

Schematic relationship between biochemistry, genetics, and molecular biology

- *Biochemistry* is the study of the chemical substances and vital processes occurring in living organisms. Biochemists focus heavily on the role, function, and structure of biomolecules. The study of the chemistry behind biological processes and the synthesis of biologically active molecules are examples of biochemistry.
- *Genetics* is the study of the effect of genetic differences on organisms. Often this can be inferred by the absence of a normal component (e.g., one gene). The study of "mutants" – organisms with a changed gene that leads to the organism being different with respect to the so-called "wild type" or normal phenotype. Genetic interactions (epistasis) can often confound simple interpretations of such "knock-out" or "knock-in" studies.
- *Molecular biology* is the study of molecular underpinnings of the process of replication, transcription and translation of the genetic material. The central dogma of molecular biology where genetic material is transcribed into RNA and then translated into protein, despite being an oversimplified picture of molecular biology, still provides a good starting point for understanding the field. This picture, however, is undergoing revision in light of emerging novel roles for RNA.
- *Chemical Biology* seeks to develop new tools based on small molecules that allow minimal perturbation of biological systems while providing detailed information about their function. Further, chemical biology employs biological systems to create non-natural hybrids between biomolecules and synthetic devices (for example

emptied viral capsids that can deliver gene therapy or drug molecules).

See also

Lists

- List of biochemistry topics
- List of biochemists
- List of biomolecules
- List of geneticists & biochemists
- List of nucleic acid simulation software
- Important publications in biochemistry (chemistry)

Related topics

- Biological psychiatry
- Biophysics
- Carbon chauvinism
- Chemical ecology
- Computational biomodeling
- EC number
- Hypothetical types of biochemistry
- International Union of Biochemistry and Molecular Biology
- Metabolome
- Metabolomics
- Molecular biology
- Molecular medicine
- Plant biochemistry
- Structural biology
- Stoichiometry
- Small molecule
- Veterinary

Notes

a. It should be noted that fructose is not the only sugar found in fruits. Glucose and sucrose are also found in varying quantities in various fruits, and indeed sometimes exceed the fructose present. For example, 32 % of the edible portion of date is glucose, compared with 23.70 % fructose and 8.20 % sucrose. However, peaches contain more sucrose (6.66 %) than they do fructose (0.93 %) or glucose (1.47 %).[6]

References

[1] Campbell, Neil A.; Brad Williamson; Robin J. Heyden (2006). *Biology: Exploring Life* (http://www.phschool.com/el_marketing.html). Boston, Massachusetts: Pearson Prentice Hall. ISBN 0-13-250882-6. .

[2] Wöhler, F. (1828). "Ueber künstliche Bildung des Harnstoffs". *Ann. Phys. Chem.* **12**: 253–256.

[3] Kauffman, G. B. and Chooljian, S.H. (2001). "Friedrich Wöhler (1800–1882), on the Bicentennial of His Birth". *The Chemical Educator* **6** (2): 121–133. doi:10.1007/s00897010444a.

[4] Ultratrace minerals. Authors: Nielsen, Forrest H. USDA, ARS Source: Modern nutrition in health and disease / editors, Maurice E. Shils ... et al.. Baltimore : Williams & Wilkins, c1999., p. 283-303. Issue Date: 1999 URI: (http://hdl.handle.net/10113/46493)

[5] Whiting, G.C (1970). "Sugars". In A.C. Hulme. *The Biochemistry of Fruits and their Products*. **Volume 1**. London & New York: Academic Press. pp. 1=31

[6] Whiting, G.C. (1970), p.5

Further reading

- Hunter, Graeme K. (2000). *Vital Forces: The Discovery of the Molecular Basis of Life*. San Diego: Academic Press. ISBN 0-12-361810-X. OCLC 162129355 191848148 44187710.

External links

- The Virtual Library of Biochemistry and Cell Biology (http://www.biochemweb.org/)
- Biochemistry, 5th ed. (http://www.ncbi.nlm.nih.gov/books/bv.fcgi?call=bv.View..ShowTOC&rid=stryer.TOC&depth=2) Full text of Berg, Tymoczko, and Stryer, courtesy of NCBI.
- Biochemistry, 2nd ed. (http://www.web.virginia.edu/Heidi/home.htm) Full text of Garrett and Grisham.
- Biochemistry Animation (http://www.1lec.com/Biochemistry/) (Narrated Flash animations.)
- SystemsX.ch - The Swiss Initiative in Systems Biology (http://www.systemsX.ch/)
- Biochemistry Online Resources (http://www.icademic.org/97445/Biochemistry/) – Lists of Biochemistry departments, websites, journals, books and reviews, employment opportunities and events.

biochemical families: proteins (amino acids/intermediates) · nucleic acids (constituents/intermediates) · carbohydrates (glycoproteins, alcohols, glycosides)

lipids (fatty acids/intermediates, phospholipids, steroids, sphingolipids, eicosanoids) · tetrapyrroles/intermediates

Article Sources and Contributors

Dual_polarization_interferometry *Source*: http://en.wikipedia.org/w/index.php?title=Dual_polarization_interferometry *Contributors*: Carlog3, Dpidurham, Elonka, Headbomb, John.parkes007, LilHelpa, Marie Poise, Materialscientist, RODERICKMOLASAR, Rjwilmsi, Sangak, Slashme, Srleffler, Stephenb, Tony1, WattsJoule, Whosasking, 38 anonymous edits

Waveguide_(optics) *Source*: http://en.wikipedia.org/w/index.php?title=Waveguide_%28optics%29 *Contributors*: Acdx, Beaumont, Bloodshedder, Bonatto, Brews ohare, Eranus, Fred138, GregorB, Gringer, Gtg204y, Hyperdeath, Igfmnbo, Inbamkumar86, Iridescent, Light current, Mild Bill Hiccup, MystRivenExile, PigFlu Oink, Radagast83, Saintrain, Sergiusz Patela, Skier Dude, Srleffler, StewartMH, The Photon, Twri, Welsh, Wmrwiki, Wtshymanski, 22 anonymous edits

Evanescent_wave *Source*: http://en.wikipedia.org/w/index.php?title=Evanescent_wave *Contributors*: Adkein, Armando12, Atkinson 291, Bryan Derksen, CardinalDan, Chinasaur, Cmdrjameson, Conscious, DARTH SIDIOUS 2, Danhash, Donarreiskoffer, Doodle77, Drdeak1, Drphilharmonic, Fgnievinski, G2star, Headbomb, Heron, Hickorybark, HortonYertleCat, Icairns, Idunno271828, Indowiz, Isidore, Jag123, Larsobrien, Laurascudder, Leroybrown, Materialscientist, MegaHasher, Merovingian, Metacomet, Meyavuz, Michael Hardy, Pediadeep, Quantum Pasty, R'n'B, RGForbes, RevRagnarok, Smoo222, Srleffler, Steve Quinn, Tarotcards, The Anome, The Photon, Thingg, Unfree, Virogtheconq, WAS 4.250, WattsJoule, Whitepaw, WikiReviewer.de, Wjbeaty, XbY, YUL89YYZ, 57 anonymous edits

Laser *Source*: http://en.wikipedia.org/w/index.php?title=Laser *Contributors*: ...adam..., .:Ajvol:., 11K, 203.37.81.xxx, 28421u2232nfenfcenc, 334a, 7Piguine, 84user, A More Perfect Onion, A-man262, A. B., A. di M., A3RO, A5, ABF, ALoopingIcon, AVand, Aaron Schulz, Abdullais4u, Abeg92, Abmcdonald, Academic Challenger, Ackatsis, Adam Martinez, Adashiel, AdjustShift, Admiral Roo, AdultSwim, Aeiouuu, Afn, Agateller, Agathoclea, AgentPeppermint, Agutie, Ahoerstemeier, Airconswitch, Alanl, Alansohn, Alberrosidus, Albrozdude, Alex-engraver, Alexius08, Alexthe5th, Alexvinny, Ali K, Ali@gwc.org.uk, Allmedia, Allstarecho, Altenmann, AmarChandra, Amh library, Amwyll Rwden, Anakin101, Ancheta Wis, Andonic, Andres, Andrew Delong, Android Mouse, Andy Dingley, Andy M. Wang, AndyZ, Angilbas, Animum, Anon user, Antandrus, Anthony, Anthony Appleyard, AntiVan, Antonio Lopez, Apricotscrub, ArchonMagnus, Arjun01, Armedtrader, Arnero, Ashenai, Ashley Pomeroy, Astrobayes, Atlant, Atropos235, Avenged Eightfold, Azkar, Azov, BDD, BK107, BOOG95, BTech United, Bacteria, Baggio10, Ballista, Banjodog, Bayerischermann, Baylink, Bbltype, Beatnik8983, Beland, Ben-Zin, BenFrantzDale, Bensaccount, Berryland1jmm, Beta34, Bhadani, Bibliomaniac15, Birge, Bkell, Blahm, Blakkandekka, Bloodshedder, Blueforce4116, Blueking12, Bobblewik, Bobo192, Bogey97, Boing! said Zebedee, Bonadea, BorisFromStockdale, Brian R Hunter, BrianWilloughby, Brianga, Brighterorange, Brigman, Brockert, Bryan Derksen, Buchanan-Hermit, Buffs, CAIRNSY90, CRGreathouse, Cajolingwilhelm, Calabraxthis, Callmarcus, Caltas, Calvinchong, Camw, Can't sleep, clown will eat me, CanadianCaesar, CanadianLinuxUser, Canderson7, Canterbury Tail, Caper13, Capricorn42, Cardamon, Carnildo, Catgut, Cats are vague, Cb6, Cbdorsett, CelticJobber, CesarB, Chaide, Chairman S., ChanningWalton, Charitwo, Charleca, Chem-awb, ChevyCha, Chewychum, Chodorkovskiy, Chovin, Chris the speller, Chrisbil09, Chrislk02, Chuck Sirloin, Chwech, Cla68, Cmdrjameson, Cmprince, Coffee, Coma28, Cometstyles, Conversion script, Corinne68, Corpx, Corrigendas, Crazyw00t1, Cremisis13, Crgibson, Criticality, Crnorizec, Ctjf83, Ctroy36, Cyktsui, Cypa, Cyphern, D367072, DV8 2XL, Daa89563, Daleious, Dan100, Danh, Danielismagic, DanielleJ, Danski14, Dante Alighieri, Davandron, Davesf, David R. Ingham, David.Mestel, David.Monniaux, Dawn Bard, Dbfirs, Dbo789, Dbtfz, Dbutch20, DeLarge, DeadEyeArrow, Deglr6328, DerHexer, Derekleungtszhei, Derfugu, Dermatonet, Dethme0w, Dhaval akbari, Dima373, Dimimimon7, Dina, Dirac66, Discosebastian, Discospinster, Dj ansi, Dlohcierekim, Dlohcierekim's sock, Dog12, Dolphonia, Domitori, Donama, DrBob, DrDentz, Dreadstar, Dreamweaver9, DrewBear11, Drhaddix, Duomillia, Dust Filter, Dysprosia, Dzenanz, DÅ,ugosz, E104421, EVula, Earendilmm, East718, Ec5618, EconomicsGuy, Ecopetition, Edebraal, Editor at Large, Edward, Eequor, Ejeffrey, El C, Elassint, Electron9, Elipongo, Ellbeecee, Ember of Light, Emilyisaac, Enviroboy, Epbr123, Equinoxe, Erc, Ergbert, Eric-Wester, Ericschulz, Espoo, Eurleif, Evanfarrar, EverGreg, Everyking, Evil Monkey, FDominec, FT2, Fabian Hassler, Falcorian, Fallenangei, Farrell0000, Fdimer, Felipe La Rotta, Femto, Fibonacci, Figureskatingfan, Fintelia, Fir0002, First Harmonic, Fishclip, Fixentries, Flakos17, FlyByPC, Flyguy649, Fourthords, Fragglet, Fram, Fredrik, Freepsbane, Freeridr, Freethron, Fresheneesz, Frodet, Fuhghettaboutit, FunPika, Furrykef, G4rfunkel, Gaius Cornelius, Gamer007, Gandhi21, Gary King, Gene Nygaard, GeoGreg, Gerkleplex, Gggrrrnade, Giftlite, Gigemag76, Gilliam, Gjd001, Gjs238, Glanhawr, Glenn, Gmankiller2, Gmaxwell, Gnomerspell, Gobonobo, Goeagles4321, Gogo Dodo, Gogobera, GossamerBliss, GraemeL, Grafen, Graham87, GrampaScience, Grand2007, Grandphuba, GreenSpigot, Greenlaser, GregorB, Grenno, Gscshoyru, Gtg204y, Guitardemon666, Gunter, Gurch, Gurchzilla, Guy M, Gwernol, HIGOCHUICHI, Hadal, Hadmack, Haham hanuka, Hairy Dude, Hamiltondaniel, Handcloud returns, Handmedown, Hankwang, HappyCamper, Happytomato, HarisM, Hawksfan18, Hayabusa future, Headbomb, HeroGiant, Herohtar, Heron, Hierzuhelfen, Hikui87, Hildenja, Hill billy wock, Hobartimus, Hooperbloob, Howcheng, Hu12, Hudsonwongzb, Hummer82793, Hut 8.5, Hydrargyrum, Hyperdeath, IW.HG, Iain99, Iamdrevil, Ian McEwen, Ianmacm, Ibakedsomecookies, Icairns, Idaltu, Igoldste, Ilikefood, Iluvjlpicard, Imacbelkin, Immachargin, Imnotminkus, Imroy, Indon, Inferno, Lord of Penguins, Infinoid, Insanity Incarnate, Instinct, Interferometrist, Iridescent, Irishguy, Isaac, Iskander32, It Is Me Here, IvanStepaniuk, Ixfd64, J.delanoy, JJLESDUDE, JLKrause, JLaTondre, JOOHD, Jabeles, Jackanape Jones, Jackaranga, James P Twomey, Janejellyroll, Janke, Janneok, Jaraalbe, Jason222, Jb.ronin, Jboyles, Jebba, Jecowa, Jellies1324, Jerome Charles Potts, Jfire, Jj137, Jknacnud, Jlaramee, Jmabel, Jmencisom, Joanjoc, JoanneB, JoeJoeseph, Joelster, Joeylawn, Joeyman365, John, John Lake, John water, John254, JohnFlux, Johncarini, Jokestress, Jonnysimpson, Josephseagullstalin, JoshuaZ, Jpkoester1, Jrockley, Julesd, Juliancolton, Junggoo, Justdaning, Jwissick, K-links, K1Bond007, Kaisershatner, Kappa, Karaboom, KathrynLybarger, Kaulike, Kbdank71, Keith D, Keith-264, Keithf@bel-fast.com, Kellen`, Kerotan, Kesac, Killermatt18, Kipkay, KittySaturn, Kkm010, Klknoles, KnowledgeOfSelf, Kntrabssi, Konor org, Koshy2000, Koyaanis Qatsi, Krash, Kristine anne, Krylonblue83, Kslays, Kungfuadam, Kuru, Kvn8907, Kyle Barbour, L337p4wn, LFW, LactoseTI, LambOfDog, Lasah, Laser-jok, Laserslight, Laservet, Laurascudder, Laurens, Lazertech, Lazylaces, LeaveSleaves, Lectonar, Lee M, Leszek Jańczuk, LibLord, Lightmouse, Lightro6, LilHelpa, Lion Info, LittleOldMe, Littlebum2002, Lord Voldemort, Loren36, Lotje, LoveMed, Luigi30, Luitgard, Luk, Lukus lee, Luna Santin, Lunamaria, Lunchscale, Lupinoid, Lysdexia, M jurrens, M-le-mot-dit, M.manary, MER-C, MPOxy, MPerel, MToolen, MacGyver07, MacedonianBoy, Macy, Mafuyu, Magister Mathematicae, Maha ts, Makele-90, Makemi, Mardavich, Mark Foskey, MarkBolton, Markand, Martarius, Martin Hedegaard, MartinHarper, Matt.T, Matthewsasse1, MattieTK, Mattyp9999, Maurice Carbonaro, Maziy300, Mbstone, McVities, Mcepjg, Measure, Medicuspetrus, Meeples, Mejor Los Indios, Mekong Bluesman, Melaen, Melongrower, Mercury, Mermaid from the Baltic Sea, Merovingian, Metaxis, Meuzzwal, Mhazard9, Miaow Miaow, Michael Hardy, Michaelbusch, Michaelpkk, Mignon, Mike R, Mike Rosoft, Mikeli88, Miquonranger03, Misos, Misskaur, Mitchandre, Mmoneypenny, Mohammad Sami, MosheA, MovementLessRestricted, Mr snarf, Mr. Lefty, Mrchrisr, Mrwojo, Muad, Muchacho Gasolino, Mufka, Muhaidib, Mulberry, Murray.booth, Mwilso24, Myanw, Mythealias, Mzajac, N734LQ, NHRHS2010, Nabla, Nagytibi, Naidevinci, Nakon, Nancy, Narayan82es, NawlinWiki, Nebrot, Neohaven, NeonDemon, Nescio, Neurolysis, Neutrality, Nevit, Niceguyedc, Nick C, Nigholith, Nikkimaria, Nivix, Nixer, Njan, Nmnogueira, Nn123645, Nneonneo, No ptr, Noclevername, Nono64, Noobdestroyah, Northumbrian, Noumenorian, Npeters07, NuncAutNunquam, Nuttycoconut, OHmanEARTHBOUND, Obnoxin, Octahedron80, Ogxela, Ohconfucius, Omegatron, Omicronpersei8, OnBeyondZebrax, Onorem, Oo7565, Opelio, Operating, Orgeat, Ose\fio, Oursroute, Owen Silverstone, Oxymoron83, PGWG, PJM, PaedosAreKewl, Palfrey, Papa November, Patrick, Patrick Berry, Paul Erik, Pawl Kennedy, PaxtonB, Peppergrower, Persian Poet Gal, Peter bertok, Phaedriel, Phil19g7, Photonicsguy, Phox63, Piano non troppo, PierreAbbat, Pmaguire, Pmg, Pmokeefe, Pmronchi, Pmurph5, Poeloq, PoolsClosed87, Portobello Prince, Potf13, Pouya, Pqrstuv, Prometheus235, Psychoticfruit, Psydev, Pzavon, QuackGuru, Quailman, Quantum ua, Quicksilvre, Quinsareth, Quintote, Quinwound, Quinxorin, Qwasty, Qwerty192837, RFNEFF, RJHall, RJaguar3, RPaschotta, Raluaces, Random user 8384993, RapidR, RattusMaximus, Ray Van De Walker, Read-write-services, Redblitz, Redhookesb, Reedy, Res2216firestar, Rettetast, RexNL, Rfwexler, Rich Farmbrough, RichMac, Richard Arthur Norton (1958-), Richbham, RiverBed, Rjwilmsi, Roadrunner, Rob Mahurin, Robin S, Rodrigja, RomanXNS, Romanm, Rory096, Rossami, Roux, RoyBoy, Rror, Rubin joseph 10, Rudolf 1922, Rudykog, Ryan, Ryan Taylor, RyanCross, S3000, SCARECROW, SDC, SEWilco, SFC9394, SG Liker, SGBailey, SJP, ST47, SWAdair, Saimhe, Salamurai, Sam nead, Samir, Sandeepjshenoy, Sanders muc, Sango123, Sapiens23, Sasuke6051, Sawt al Hurriyah, Sbharris, Sceptre, SchfiftyThree, Schnazola, Schutz, Scooteemooty, Scott Paeth, Scray, Sct72, SeanMack, Seanweir, Searchme, Seba5618, Seeker alpha806, Sgeo, Shaddack, Shadow148, Shawn K. Quinn, Shawn Worthington Laser Plasma, Shinpah1, Shirik, Shoeofdeath, Shoessss, SiPlus, Sietse Snel, Silly rabbit, Simesa, Singularity, SkerHawx, Sladen, Slashme, Slavik81, Sleigh, Slgrandson, Smartkid23, SmileToday, Smithbrenon, Snacky, SnoKoneManiac, Snowboy9999, Snowolf, Soliloquial, Sonett72, Sorset, Sosodank, Spangineer, Special needs LOL, SpectrumDT, SpookyMulder, Sportzplyr9090, Srleffler, Srnec, StealthFox, Stefan Udrea, Stephenb, Steve, Stevertigo, Stonejag, Stratocracy, Styrofoam1994, Super Rad!, Superm401, SusikMkr, Switchercat, Swpb, Symane, Sysy, TKD, Talaga87, Tapir Terrific, Tarchon, Taroaldo, Taune, Tazzaler, TedE, TenOfAllTrades, The Font, The Magnificent Clean-keeper, The Photon, The Rambling Man, The wub, TheObtuseAngleOfDoom, TheOtherSiguy, TheParanoidOne, Theresa knott, Thingg, Thinghy, Think outside the box, Thomas Paine1776, Thorpe, ThreeDee912, Thumperward, Tillman, Tim Starling, Tim1988, TimProof, TimVickers, Time3000, Timo Honkasalo, Tiptoety, Titoxd, Tlabshier, Tobby72, Todd Vierling, Tom Duff, Tomchiukc, Tomer yaffe, Tony Fox, Tony1, Tpbradbury, Tresiden, Treuzgirl16, Truthflux, Tslocum, Ttiotsw, TurnedWorm, Twice25, USS Noob Hunter, UberScienceNerd, Udayan.choudhury, Udzu, Ukexpat, Ultraexactzz, UserDoe, Utcursch, Valkyryn, Vanished User 8a9b4725f8376, Vary, Vinsm, Vipinmathew, Vishnava, Vlizzardv, Waggers, WakeUp, Walter Humala, Wapcaplet, Warharmer, Wavelength, Waxigloo, Wcleveland, Wehttam1106, Weihao.chiu, Welsh21, Wenli, Whkoh, Wiggydo, Wiki fanatic, Wikid77, Wikifan798, Wikipediarules2221, Willking1979, Withead, Wizard191, Wjbeaty, Womza, Woohookitty, Work permit, Worldwalker, Wyatt915, XLostMemoriesx, Xanzzibar, Xezbeth, Xiahou, Xp54321, XxXXMULLIGANXXxx, Yachtsman1, Yath, Yatoast, Yosri, Youngjim, Zaledin, Zaphraud, ZeframCochrane, Zimbres, Zotel, Zundark, BOΥ, Σ, سمرقندی, , , 1772 anonymous edits

Conformational_change *Source*: http://en.wikipedia.org/w/index.php?title=Conformational_change *Contributors*: AJim, Alexbateman, Ark2120, Carstensen, Chris the speller, Dcirovic, Deb, Fqreunyg352, Hannes Röst, Hodja Nasreddin, Kane5187, Kilva, Malcolma, Nbanaval, Paul P, R'n'B, Salafsky, Squirrel teeth, Stemonitis, 8 anonymous edits

Conformation_activity_relationship *Source*: http://en.wikipedia.org/w/index.php?title=Conformation_activity_relationship *Contributors*: EdChem, Gronan, Hodja Nasreddin, Malcolma, Materialscientist, Nick Number, RHaworth, Rich Farmbrough, The-red-dragon-keeper, 3 anonymous edits

Interactome *Source*: http://en.wikipedia.org/w/index.php?title=Interactome *Contributors*: Aejohnst, Alexbateman, AlistairMcMillan, AnteaterZot, Arjayay, Ashgene, Augustulus2, Billgordon1099, Ceyockey, CopperKettle, Hodja Nasreddin, Jebus989, Jesse V., Jheppner, Jong, Jongbhak, Kkmurray, Lantonov, Maja3141, Minnsurfur2, Nagelfar, Nardojacq, Navratil vincent, Niteowlneils, Peteruetz, Ph.eyes, PhilipNRees, Pindb, Plindenbaum, Rajah, Rholton, Rjwilmsi, Rofl, Saint Aardvark, Shinwachi, Shureg, T4exanadu, Teledildonix314, Thedjatclubrock, Thorwald, Tony1, Vimes656, 58 anonymous edits

Biochemistry *Source*: http://en.wikipedia.org/w/index.php?title=Biochemistry *Contributors*: 168..., 203.20.98.xxx, 217.228.14.xxx, 2pac 2007, 52G, A8UDI, APH, Abductive, AdamRetchless, Ahoerstemeier, Alaadabarita, Alai, Alansohn, Alcarreau, Ale jrb, Alex '05, Alexandria, Allstarecho, Amaher, Amitashsam, Andre Engels, AndreasJS, Andreworkney, Andris, Anpetu-We,

Antandrus, Applepie390, Arakunem, Arcadian, Aremith, Arifur6051, Arthena, Ash421, Avenged Eightfold, AxelBoldt, Babyboy004, Barefootguy, Basilicofresco, Bduke, Ben Zealley, BenB4, Bennetto, Bensaccount, Bettia, Blainster, Bluemask, Bob, Bobo192, Bonadea, Boothy443, Booyabazooka, Bornslippy, BostonBiochem, Brandonrush, Bryan Derksen, Bsadowski1, Bubba hotep, COMPFUNK2, Cabe6403, Calliopejen1, Caltechdoc, CambridgeBayWeather, Canuckian89, Capricorn42, Card, Carstensen, Caulde, Cbailey7, Ceyockey, ChemNerd, Chika11, Chris jf, Chris the speller, Christian75, Chriswiki, Chun-hian, ClockworkSoul, Cometstyles, Confiteordeo, Conversion script, Coppertwig, Courcelles, Curb Chain, Curps, Cyp, D, DGG, DJ Clayworth, DPistotallyawesome, Dan East, Danski14, Darth Panda, Deafgirl, Defeatedfear, Dekisugi, Deor, DerHexer, Diberri, Discospinster, Djramone, Dmb000006, Doctorbruno, Donarreiskoffer, Doug swisher, Doulos Christos, Dramagirlforjesus, Dreadstar, Drmies, Drphilharmonic, Dwmyers, Edgar181, El C, Enchanter, Epbr123, Erik9, Ermysted's school, Espresso Addict, Essam Sharaf, EtherROR, Etxrge, Excirial, Explosivedeath7, Fakih1234, Fastily, Fathimaazara, Favonian, Figma, Firewall62, Flyguy649, Frankenpuppy, Fvandrog, Gamewizard71, Gareth Jones, Gentgeen, Giftlite, Gilliam, Giraffedata, Glenn, Gludwiczak, Go for it!, GoShow, GoodDamon, Grafen, GrahamColm, Gscshoyru, Guanaco, Gumball456, H Padleckas, H.tjerns, Haham hanuka, Halcatalyst, Headbomb, HenryLi, HiDrNick, Hydrogen Iodide, I already forgot, I dream of horses, Into The Fray, Itub, J Di, J. Finkelstein, J. M., J.delanoy, JRHorse, Jag123, Jamesscottbrown, Jauhienij, Jcsvarela, Jdcollins13, Jeff G., Jenvargas93, JeramieHicks, JeremyA, Jethero, Jhay116, Jim1138, Jobseeker-us, John Lake, Jotomicron, Joud98, Jovianeye, Justacontributor, Jytdog, K.murphy, KLassetter, Kandar, Karol Langner, Kbdank71, Keilana, Kevyn, Kipala, Kiwakwok, Kjaergaard, Knowledge Seeker, Knutux, Kocio, Kowey, Krellis, Kuru, L Kensington, Lalsingh, Leafyplant, Lexor, Lfh, Lightmouse, Lil'swallowtail, LilHelpa, Lineweaver, Linnea, Logan baum, Loka1282, Loohcsnuf, Looper5920, Loren.wilton, Luk, M stone, MER-C, Mac, Magnus Manske, Mah159, Majorly, Mani1, MarcoTolo, Marj Tiefert, Marskell, Marywhi, Mattbrundage, Maurreen, Mav, Mayooranathan, Mayumashu, McGeddon, Mendalus, Mets501, Mikael Häggström, Mike Peel, Mikez, Mikhail Garouznov, Mishlai, Monty845, Moreschi, MrOllie, Mushin, Mxn, N4nojohn, N8mills, Naturespace, NawlinWiki, Neilmitch02, Netoholic, Neuroticopia, Neverquick, Night of the Big Wind, Nighttoy, Notinasnaid, Nssr 84, Ntropy25, Occhanikov, Ohsoserious, OldakQuill, Ombudsman, Orioneight, Ornil, Oxymoron83, P99am, PM800, Palica, Passw0rd, Paul-L, Paulkappelle, Pedro, Persian Poet Gal, Peterlin, Pgk, Pi, PiFanatic, Pinethicket, Pmrich, Potaaatos, Promage281, Pschemp, Pscott22, Pvosta, Quadalpha, Quaeler, Quintote, Qwertyman33, Ragesoss, RaseaC, RebelBodhi, Remi0o, Rettetast, RexNL, Rich.buckman, Richard001, Rifleman 82, Riomack, Rob Hooft, Rodelapa, Rominandreu, Rrburke, Sadi Carnot, Salvaje20, Samsara, Sarakoth, Sbharris, Scorpion451, ScottJ, Sdneidich, Serein (renamed because of SUL), Shadowjams, Shirulashem, Smalljim, Snowmanradio, Someone else, Squidonius, Staib, Steinsky, Stephenb, Stevendboy, StevieNic, Stewartadcock, Stupid2, SunCreator, Superbeecat, Tanaats, Tarcieri, Tarquin, Tbeyett, Tcncv, Techneaux, Tedder, Template namespace initialisation script, Templatehater, Terebigemuwan, The Thing That Should Not Be, The Transhumanist, The Transhumanist (AWB), TheAlphaWolf, Thecheesykid, Themfromspace, Theruchet, Thomaswgc, ThyCantabrigde, Tiddly Tom, Tide rolls, Tiggerjay, TimBuck2, TimVickers, Timir2, Tobby72, Tobias Conradi, Tobias Hoevekamp, Tom the biochemist, Trainra, Twas Now, UnitedStatesian, Untitled300, Vincent kraeutler, Vivin, Vsmith, Washburnmav, Weewengweng, WereSpielChequers, WikipedianMarlith, Willisis2, Wimt, Winkel111, Wknight94, Wotnow, Wouterstomp, WriterHound, X201, YannisKollias, Yikrazuul, Youandme, Zeb edee, Zephyris, °¡°, Александър, 696 anonymous edits

Image Sources, Licenses and Contributors

Image:slab waveguide 1.svg *Source*: http://en.wikipedia.org/w/index.php?title=File:Slab_waveguide_1.svg *License*: unknown *Contributors*: SkiDragon, The Photon

Image:Optical-fibre.svg *Source*: http://en.wikipedia.org/w/index.php?title=File:Optical-fibre.svg *License*: unknown *Contributors*: User:Gringer

Image:Electron density wave - plasmon excitations.png *Source*: http://en.wikipedia.org/w/index.php?title=File:Electron_density_wave_-_plasmon_excitations.png *License*: unknown *Contributors*: Center for Probing the Nanoscale -

File:Total internal reflection.jpg *Source*: http://en.wikipedia.org/w/index.php?title=File:Total_internal_reflection.jpg *License*: unknown *Contributors*: User:Fir0002

File:Evanescent wave.jpg *Source*: http://en.wikipedia.org/w/index.php?title=File:Evanescent_wave.jpg *License*: unknown *Contributors*: Aluminium

File:Military laser experiment.jpg *Source*: http://en.wikipedia.org/w/index.php?title=File:Military_laser_experiment.jpg *License*: unknown *Contributors*: US Air Force

File:RGB_laser.jpg *Source*: http://en.wikipedia.org/w/index.php?title=File:RGB_laser.jpg *License*: unknown *Contributors*: User:BOT

File:Laser play.jpg *Source*: http://en.wikipedia.org/w/index.php?title=File:Laser_play.jpg *License*: unknown *Contributors*: Jeff Keyzer from San Francisco, CA, USA

File:Laser.svg *Source*: http://en.wikipedia.org/w/index.php?title=File:Laser.svg *License*: unknown *Contributors*: User:Tatoute

File:Stimulated Emission.svg *Source*: http://en.wikipedia.org/w/index.php?title=File:Stimulated_Emission.svg *License*: unknown *Contributors*: User:V1adis1av

File:Laser DSC09088.JPG *Source*: http://en.wikipedia.org/w/index.php?title=File:Laser_DSC09088.JPG *License*: unknown *Contributors*: Copyright © 2004 David Monniaux

File:Helium neon laser spectrum.svg *Source*: http://en.wikipedia.org/w/index.php?title=File:Helium_neon_laser_spectrum.svg *License*: unknown *Contributors*: User:Papa November

File:Aleksandr Prokhorov.jpg *Source*: http://en.wikipedia.org/w/index.php?title=File:Aleksandr_Prokhorov.jpg *License*: unknown *Contributors*: Nobel foundation

File:Gould notebook 001.jpg *Source*: http://en.wikipedia.org/w/index.php?title=File:Gould_notebook_001.jpg *License*: unknown *Contributors*: User:Srleffler

File:History of laser intensity.svg *Source*: http://en.wikipedia.org/w/index.php?title=File:History_of_laser_intensity.svg *License*: unknown *Contributors*: ArcyQwerty, Grendelkhan, Rugby471, Slashme, 1 anonymous edits

File:Commercial laser lines.svg *Source*: http://en.wikipedia.org/w/index.php?title=File:Commercial_laser_lines.svg *License*: unknown *Contributors*: Danh, 2 anonymous edits

File:Green-laser-pointer-dpss-diagrams.jpg *Source*: http://en.wikipedia.org/w/index.php?title=File:Green-laser-pointer-dpss-diagrams.jpg *License*: unknown *Contributors*: User:Laser Chris

File:Diode laser.jpg *Source*: http://en.wikipedia.org/w/index.php?title=File:Diode_laser.jpg *License*: unknown *Contributors*: Original uploader was Deglr6328 at en.wikipedia

File:Laser sizes.jpg *Source*: http://en.wikipedia.org/w/index.php?title=File:Laser_sizes.jpg *License*: unknown *Contributors*: Maksim, Shyam, 1 anonymous edits

File:Lying down on the VLT platform.jpg *Source*: http://en.wikipedia.org/w/index.php?title=File:Lying_down_on_the_VLT_platform.jpg *License*: unknown *Contributors*: G. Hüdepohl/ESO

File:DIN 4844-2 Warnung vor Laserstrahl D-W010.svg *Source*: http://en.wikipedia.org/w/index.php?title=File:DIN_4844-2_Warnung_vor_Laserstrahl_D-W010.svg *License*: unknown *Contributors*: User:DrTorstenHenning

File:Laser label 2.jpg *Source*: http://en.wikipedia.org/w/index.php?title=File:Laser_label_2.jpg *License*: unknown *Contributors*: User:BOT

File:Network of how 100 of the 528 genes identified with significant differential expression relate to DISC1 and its core interactors.png *Source*: http://en.wikipedia.org/w/index.php?title=File:Network_of_how_100_of_the_528_genes_identified_with_significant_differential_expression_relate_to_DISC1_and_its_core_interactors.png *License*: unknown *Contributors*: Hennah, Porteous

File:Gerty Theresa Radnitz Cori (1896-1957) and Carl Ferdinand Cori.jpg *Source*: http://en.wikipedia.org/w/index.php?title=File:Gerty_Theresa_Radnitz_Cori_(1896-1957)_and_Carl_Ferdinand_Cori.jpg *License*: unknown *Contributors*: Smithsonian Institution from United States

Image:Sucrose-inkscape.svg *Source*: http://en.wikipedia.org/w/index.php?title=File:Sucrose-inkscape.svg *License*: unknown *Contributors*: Don A. Carlson

Image:Fat triglyceride shorthand formula.PNG *Source*: http://en.wikipedia.org/w/index.php?title=File:Fat_triglyceride_shorthand_formula.PNG *License*: unknown *Contributors*: Wolfgang Schaefer

Image:AminoAcidball.svg *Source*: http://en.wikipedia.org/w/index.php?title=File:AminoAcidball.svg *License*: unknown *Contributors*: Edgar181, TimVickers, Wutsje, YassineMrabet, 2 anonymous edits

Image:DNA chemical structure.svg *Source*: http://en.wikipedia.org/w/index.php?title=File:DNA_chemical_structure.svg *License*: unknown *Contributors*: Hystrix, Madprime, Wickey, 5 anonymous edits

Image:Beta-D-Glucose.svg *Source*: http://en.wikipedia.org/w/index.php?title=File:Beta-D-Glucose.svg *License*: unknown *Contributors*: User:Yikrazuul

Image:Saccharose.svg *Source*: http://en.wikipedia.org/w/index.php?title=File:Saccharose.svg *License*: unknown *Contributors*: Booyabazooka, Calvero, Edgar181, 1 anonymous edits

Image:Cellulose-2D-skeletal.png *Source*: http://en.wikipedia.org/w/index.php?title=File:Cellulose-2D-skeletal.png *License*: unknown *Contributors*: Benjah-bmm27, Edgar181, Slashme

Image:1GZX Haemoglobin.png *Source*: http://en.wikipedia.org/w/index.php?title=File:1GZX_Haemoglobin.png *License*: unknown *Contributors*: Original uploader was Zephyris at en.wikipedia

Image:Amino acids 1.png *Source*: http://en.wikipedia.org/w/index.php?title=File:Amino_acids_1.png *License*: unknown *Contributors*: Arria Belli, CommonsDelinker, CommonsDelinkerHelper, Edgar181, Karelj, Matanya (usurped), OsamaK, 1 anonymous edits

Image:Schematic relationship between biochemistry, genetics and molecular biology.svg *Source*: http://en.wikipedia.org/w/index.php?title=File:Schematic_relationship_between_biochemistry,_genetics_and_molecular_biology.svg *License*: unknown *Contributors*: OldakQuill, PatríciaR

MIX
Papier aus verantwortungsvollen Quellen
Paper from responsible sources
FSC® C105338

Printed by Books on Demand GmbH, Norderstedt / Germany